Managing Postharvest Quality and Losses in Horticultural Crops

The Editors

Dr. K.L. Chadha is currently the President of The Horticultural Society of India as also Adjunct Professor (Hort.) at IARI. He received his B.Sc. (Agri) and M.Sc. (Hort.) degrees from Punjab University, and Ph.D. from Indian Agricultural Research Institute (IARI), New Delhi.

He has served in various capacities at IARI, Punjab Agricultural University, Ludhiana Central Mango Research Station, Lucknow. He was the Director, IIHR, Bangalore, Horticulture Commissioner and Executive Director, National Horticulture Board, Govt. of India; Deputy Director General (Hort.), ICAR and National Professor (Hort.) at IARI.

He has authored/edited nearly 4 dozen books and bulletins including the series "Advances in Horticulture" published in 13 volumes covering 9,410 pages. He has been associated with several scientific societies as their Chief Patron, President or Vice President. He has served as consultant to several national and international organisations.

Dr. Chadha is the recipient of several awards for his outstanding contributions to R & D in horticulture. He is a Fellow of NAAS, New Delhi and ISHS, Belgium and D.Sc. (h.c.) from three State Agricultural Universities. Some important awards won by him include Borlaug Award, Om Prakash Bhasin Award, Dr. H.M. Marigowda Award, S.K. Mitra Memorial Award, S.S. Ranade Memorial Life Time Achievement Award, HSI-Shivashakthi Life-Time Achievement Award, B.P. Pal Memorial Award, National Agriculture Leadership Award and Krishi Shiromanni Samman.

Dr K.L. Chadha has steered policy planning, research and development of horticulture in India for almost five decades. For his yeomen's service Dr Chadha was conferred the prestigious *Padma Shri* by Her Excellency, the President of India.

Prof. R. K. Pal is currently working as the Director of the ICAR-National Research Center on Pomegranate at Solapur, Maharashtra. After his formal education from West Bengal, he received M.Sc. and Ph.D. degrees in Horticulture from IARI and joined Agricultural Research Service of ICAR. He served IARI in different capacities for over 25 years as Scientist, Professor and Head of the Division of PHT. He underwent advanced training in USA and attended the International Course on Food Processing as FAO Fellow in the Netherlands. He has published more than 100 research/technical papers in national and international journals and guided several research scholars at IARI, New Delhi currently occupying prestigious positions in India, USA, Iran, Vietnam and in many other countries. He is a Fellow of Horticultural Society of India, and has won Dr. J.C. Anand Gold Medal for his contribution in PHT of horticultural crops. He was a national consultant of FAO, member of the Technical Committee of SAARC and Technical Consultant for FHEL-operated CA store. He represented IARI as a core committee member for establishment of Advance Centre for Agriculture Research and Education in Myanmar.

Managing Postharvest Quality and Losses in Horticultural Crops

Volume 1
General Issues

Editors

K.L. Chadha

President, The Horticulture Society of India,
F-1 Block, NASC Complex, DPS Marg, Todapur,
New Delhi – 110 012, India

and

R.K. Pal

Director, ICAR-National Research Centre on Pomegranate
NH-9, Solapur-Pune Highway, Kegaon,
Solapur – 413 255, Maharashtra, India

2019

Daya Publishing House®

A Division of

Astral International Pvt. Ltd.

New Delhi – 110 002

Managing Postharvest Quality and Losses in Horticultural Crops (3 Volume Set)

Volume 1: General Issues	Page 001-251
Volume 2: Fruit Crops	Page 253-579
Volume 3: Vegetables, Flowers & Plantation Crops	Page 581-755

First Published, 2015

ISBN 978-93-5124-355-7 (Vol.1)
ISBN 978-93-5124-358-8 (Set)

Published by : **Daya Publishing House®**
A Division of
Astral International Pvt. Ltd.
– ISO 9001:2015 Certified Company –
4736/23, Ansari Road, Darya Ganj
New Delhi-110 002
Ph. 011-43549197, 23278134
E-mail: info@astralint.com
Website: www.astralint.com

Laser Typesetting : **Classic Computer Services**, Delhi - 110 035

Printed at : **Replika Press Pvt. Ltd.**

PRINTED IN INDIA

Prof. M. S. Swaminathan
Founder Chairman, M S Swaminathan Research Foundation
Third Cross Street, Taramani Institutional Area
Chennai – 600 113 (India)
Email: swami@mssrf.res.in

MSSRF

Foreword

During 1981 the Government of India had constituted a committee under my chairmanship to suggest a scientific approach to postharvest management of perishables in the country. Among various recommendations made, the Committee suggested the establishment of a National Horticulture Board to lay emphasis on programmes related to Postharvest Management. The Committee also identified a group of 12 horticultural crops covering fruits, vegetables and spices for postharvest research apart from several other new crops that were gaining commercial importance. Among fruits apple, mango, banana, citrus, grape, guava, papaya, pineapple, pomegranate and sapota were identified. While among vegetables and spice crops commercial crops like onion, potato, pepper/chilli, turmeric and ginger were included. The National Horticulture Board of the Ministry of Agriculture with its headquarters at Gurgaon is an important outcome of the recommendations.

Even though research on Postharvest Management has been underway in India at the Central Food Technological Research Institute (CFTRI), Mysore, it was not adequate to cover all major crops and aspects. Thus ICAR as a nodal organization implemented a project on Postharvest Management under PL-480 scheme and subsequently set up the Central Institute for Post Harvest Engineering and Technology at Ludhiana. In addition, Defence Food Research Laboratory (DFRL), Indian Institute of Crop Processing Technologies (IICPT) Thanjavur, Indian Institute of Technology (IIT) Kharagpur and Mumbai, Jhadavpur University, Kolkata, Bhaba Atomic Research Centre (BARC), Mumbai and many other private funded Institutions have been giving emphasis on these aspects. Of late, new institutions *viz.* National Institute of Food Technology Entrepreneurship and Management (NIFTEM) at Kondili, Haryana and National Agri-Food Biotechnology Institute (NABI), Mohali, Punjab have also been established in view of the emergent need for combating the huge postharvest losses of perishable commodities taking place in India. As a result several technologies have been developed on postharvest management of perishable horticultural crops by these institutions. However, most of these technologies have not been transferred to the end users in an appropriate manner due to non-availability of recommendations in an implementable form. Further, all these recommendations are not available at one place and are mostly concealed in research papers or reports which are neither available nor comprehensive.

I am happy that Dr. K.L. Chadha, former Deputy Director General (Horticulture), ICAR and Dr. R.K. Pal, Director, ICAR-National Research Centre on Pomegranate, Solapur, Maharashtra have made an excellent effort in consolidating the scattered information available from various source, compiled it and edited it in a lucid and understandable format in this book titled "Managing Postharvest Quality and Losses in Horticultural Crops" with conclusive recommendations for the benefit of different stake holders. I congratulate the editors for this effort at a time when the horticulture production has overtaken the food crop production and its post-harvest management warrants the use of scientific technologies.

I hope this book will be of useful to all engaged in wholesale, retail and export trade of horticultural crops as well as researchers and students. It is also hoped that researchers will make further efforts to plug the gaps in technologies wherever these exist. At the same time, we should continue to make effort to bridge the gap between scientific know-how and field level do-how.

M.S. Swaminathan

Preface

India is the second largest producer of fruits and vegetables in the world but it encounters huge postharvest losses almost equivalent to the total production of horticultue crops of the United Kingdom. Reduction in postharvest losses of horticultural crops is less expensive than bringing equivalent increase in production. Postharvest research in India was started at CFTRI, Mysore. However, with the increase in production and productivity of horticultural crops in India, ICAR launched a PL-480 scheme with the assistance of the United States of America to undertake research on postharvest technology of horticultural crops at IARI, New Delhi as lead centre. Later on an All India Coordinated Research Project on Postharvest Technology of Horticultural Crops was launched with a few centres at ICAR institutes and state agricultural universities. During the last three decades or so, lot of information on several aspects of postharvest management of horticultural crops has been generated in India. The horticulture industry in India has grown leaps and bounds with the increase in area under cultivation, production and export. However, scienitific postharvest management and leading adoption of international standards for export quality of horticultural produce from India with respect to combating various tariff and non-tariff barriers is still in its infancy. This book has compiled information keeping in view the needs of horticultural entreprenerus who would like to take-up domestic and export trade of fruits and vegetables on scientific basis and do not have access to this vital information.

The book is divided into three volumes comprising of 1) General Issues, 2) Postharvest Management of Fruit Crops and 3) Postharvest Management of Vegetables, Flowers and Plantation Crops. Eminent research workers who have been engaged in the area of Postharvest Management in the NARS have contributed chapters in this publication. Efforts have been made to incorporate information on status of cultivation, cultivars, association between pre-harvest crop management practices and quality after harvest, pack-house operations, storage and marketing for domestic and export trade. The book is also suitably illustrated with colour photographs, wherever necessary. Important references to each subject have been cited for detailed reading. The authors acknowledge the compilation of chapters using scattered information that is available in public domain in the form of research papers and in relevent websites.

The editors are grateful to all the contributors for their cooperation, support and timely submission of their manuscripts for bringing out this publication. The editors also express thanks to Mrs. Mandira Pal for her active support in correcting proofs of the book. The cooperation and support of the staff of the Horticultural Society of India especially Mr. Santosh Kr. Atre is duly acknowledged. The editors are extremely grateful to Prof. M.S. Swaminathan for agreeing to write the Foreword of the book.

It is hoped that this publication will be of immense value to all the stakeholders *viz*. horticultural entrepreneurs, retailers, wholesalers and exporters engaged in the business of supply chain of perishable horticultural produce, research scholars engaged in development of postharvest management protocols, and others in view of the objectives for which it has been written, *i.e.,* reducing postharvest losses and delivering quality produce both in domestic and international market.

K.L. Chadha

R.K. Pal

Contents

Foreword *v*

Preface *vii*

List of Contributors *xiii*

Volume 1: General Issues

1. **Postharvest Management: An Overview** **1**
K.L. Chadha and R.K. Pal

2. **Fruit Bagging** **31**
Ram Asrey

3. **Maturity Indices** **41**
R.K. Pal and P. Suresh Kumar

4. **Mechanisation in Harvesting** **77**
S.C. Mandhar and R.T. Patil

5. **Pre and Postharvest Management** **89**
B.V.C. Mahajan and Alemwati Pongener

6. **Postharvest Diseases** **111**
R.D. Rawal and Jyotsana Sharma

7. **Grading and Packaging** 147
M.S. Ladaniya

8. **Insect Disinfection and Quarantine** 211
Abraham Verghese and M.A. Rashmi

9. **Pesticide Residues** 231
Ahammed Shabeer T.P. and Kaushik Banerjee

Volume 2: Fruit Crops

10. **Almond** 253
Desh Beer Singh and Nazeer Ahmed

11. **Apple** 269
R.R. Sharma and R.K. Pal

12. **Banana** 291
C.K. Narayana

13. **Ber** 319
Sunil Pareek

14. **Citrus Fruits** 333
D.V. Sudhakar Rao and M.S. Ladaniya

15. **Custard Apple** 361
Sunil Pareek

16. **Date Palm** 381
Sunil Pareek

17. **Grape** 407
Ajay Kumar Sharma and P.G. Adsule

18. **Guava** 423
H. Ravishankar, A.K. Singh and Neelima Garg

19. **Litchi** 441
Vishal Nath and S.K. Purbey

20. **Mango** 461
W.S. Dhillon

21. **Papaya** 485
V. Ponnuswamy and T.L. Preethi

22. Pineapple 497
Bidyut C. Deka

23. Pomegranate 517
R.K. Pal and K. Dhinesh Babu

24. Sapota 535
Dev Raj and Ashok K. Senapati

25. Strawberry 549
Ram Asrey

26. Walnut 561
Desh Beer Singh and Nazeer Ahmed

Volume 3: Vegetables, Flowers and Plantation Crops

27. Cashew 581
D. Balasubramanian and P.L. Saroj

28. Chilli 597
Surya Kumari Surepeddi and Giridhar Kalidasu

29. Cut Flowers 617
Kushal Singh and Ramesh Kumar

30. Garlic 643
K.E. Lawande and P.C. Tripathi

31. Ginger 655
Ananta Saikia and Sanjib Sharma

32. Oil Palm 677
S. Arulraj

33. Onion 691
R.P. Gupta and H.P. Sharma

34. Orchids 709
L.C. De and N. Pathak

35. Potato 727
R. Ezekiel

List of Contributors
Vol. 1

Ahammed Shabeer, T.P. (Dr.)
Scientist, ICAR-National Research Centre for Grapes, P.O. No.3, Manjri Farm, Pune – 412 307, Maharashtra, India
E-mail: shabsnrcg@gmail.com

Asrey, Ram (Dr.)
Principal Scientist, Division of Food Science & Postharvest Technology, ICAR-Indian Agricultural Research Institute, New Delhi – 110 012, India
E-mail: ramu_211@yahoo.com

Banerjee, Kaushik (Dr.)
National Fellow, ICAR-National Research Centre for Grapes, P.O. No.3, Manjri Farm, Pune – 412 307, Maharashtra, India
E-mail: kbgrape@yahoo.com

Chadha, K.L. (Dr.)
President, The Horticulture Society of India, F-1 Block, NASC Complex, DPS Marg, Todapur, New Delhi – 110 012, India
E-mail: klchadha@gmail.com

Ladaniya, M.S. (Dr.)
Director, ICAR-National Research Centre for Citrus, Amaravati Road, Shankar Nagar Post Office, Kachimet, Nagpur – 440 010, Maharashtra, India
E-mail: msladaniya@gmail.com

Mahajan, B.V.C. (Dr.)
Professor, Punjab Horticultural Postharvest Technology Centre, Punjab Agricultural University, Ludhiana – 141 004, Punjab, India
E-mail: bvc_mahajan@rediffmail.com

Mandhar, S.C. (Dr.)
Emeritus Scientist, ICAR-Indian Institute of Horticultural Research, Hessaraghatta, Bengaluru – 560 089, Karnataka, India
E-mail: suresh.mandhar@gmail.com

Pal, R.K. (Dr.)
Director, ICAR-National Research Centre on Pomegranate, NH-9, Solapur-Pune Highway, Kegaon, Solapur – 413 255, Maharashtra, India
E-mail: rkrishnapal@gmail.com

Patil, R.T. (Dr.)
Former Director, ICAR-Central Institute of Postharvest Engineering and Technology, P.O. PAU, Ludhiana – 141 004, Punjab, India
E-mail: ramabhau@gmail.com

Pongener, Alemwati (Dr.)
Scientist, ICAR-National Research Centre on Litchi, Mushahari, P O Ramna, Muzaffarpur – 842 002, Bihar, India
E-mail: alemwati@gmail.com

Rashmi, M.A. (Dr.)
Scientist, C/o Department of Entomology, University of Agricultural Sciences, GKVK, Bengaluru – 560 065, Karnataka, India
E-mail: rashmigowda.ento@gmail.com

Rawal, R.D. (Dr.)
Former Head, Division of Plant Pathology, ICAR-Indian Institute of Horticultural Research, Hessaraghatta, Bengaluru – 560 089, Karnataka, India
E-mail: rawal.rd@gmail.com

Sharma, Jyotsana (Dr.)
Principal Scientist, ICAR-National Research Centre on Pomegranate, NH-9, Solapur-Pune Highway, Kegaon, Solapur – 413 255, Maharashtra, India
E-mail: jyotisharma128@yahoo.com

Suresh Kumar, P. (Dr.)
Senior Scientist, ICAR-National Institute of Abiotic Stress Management Malegaon, Baramati – 413 115, Maharashtra, India
E-mail: psureshars@gmail.com

Verghese, Abraham (Dr.)
Director, ICAR-National Bureau of Agriculturally Important Insects, Post Box No. 2491, H.A. Farm, Post Bellary Road, Bengaluru – 560 024, Karnataka, India
E-mail: avergis@iihr.ernet.in

Chapter 1

Postharvest Management: An Overview

K.L. Chadha
President, The Horticulture Society of India,
F-1 Block, NASC Complex, DPS Marg, Todapur,
New Delhi – 110 012, India
E-mail: klchadha@gmail.com

R.K. Pal
Director, ICAR-National Research Centre on Pomegranate,
NH-9, Solapur-Pune Highway, Kegaon,
Solapur – 413 255, Maharashtra, India
E-mail: rkrishnapal@gmail.com

1.0 Introduction

The horticultural crops comprise of a wide range of fruits, vegetables, spices, flowers, mushroom, bamboo, plantation crops and non food crops like medicinal and aromatic plants. Three decades back the role of these crops was realized in India through cropping diversification as a remunerative, viable and sustainable alternative to rice-wheat cropping system and their comparative advantage in livelihood security. As a result, policy attention to horticulture has been given since VII Five Year Plan. Emphasis on horticulture crops has resulted in higher growth rate of 5 per cent compared to 2.5 per cent from agriculture. Horticulture also contributes 30 per cent of GDP compared to 17 per cent from agriculture. It also enjoys 52 per cent share of exports from 15 per cent cultivated area. As a result of this, the area, production and productivity of horticultural crops has increased significantly. Production of horticultural crops in India increased from 94.07 million tonnes in 1991-92 to 268.8

million tonnes in 2012-13. It is now fully recognized that in order to achieve a growth rate of 4 per cent in agriculture, horticulture has to grow at a much faster and higher rate.

1.1 Nutritional Benefits

Fruits and vegetables are vital to improve the nutritional quality of our diet: together these supply not only vitamins and minerals but also the much needed roughage; at the same time these are low on calorie count. Fruits and vegetables are sometimes called as powerhouse of antioxidants having anti-ageing and anti cancer properties. Vitamin C, E, carotenoids (which provide fruits and vegetables their natural colour) and mineral selenium are powerful antioxidants. Fruits and vegetables also help to maintain a healthy gastrointestinal tract and a better vision. Some of the potential sources of antioxidants and their health benefits are mentioned in Table 1.

Table 1: Health Benefits of Antioxidants and their Food Sources

Antioxidant	*Health Benefits*	*Food Sources*
β-carotene	Keeps skin healthy, helps prevent night blindness and infections, promotes growth and bone development.	Red, yellow-orange, and leafy green vegetables and fruits, including carrot, apricot, cantaloupe, pepper, tomato, spinach, broccoli, sweet potato, and pumpkin.
Selenium	Helps maintain healthy hair and nails, enhances immunity, works with vitamin E to protect cells from damage. Reduces the risk of cancer, particularly lung, prostate, and colorectal.	Garlic, seeds and Brazil nuts.
Vitamin C	Destroys free radicals inside and outside cells. Helps in the formation of connective tissue, the healing of wounds, and iron absorption; also helps to prevent bruising and keep gums healthy. May reduce risk of cataracts, heart disease, and cancer.	Pepper, tomato, citrus fruits and juices, berries, broccoli, spinach, cabbage, potato, mango and papaya.
Vitamin E	Acts as the protector of essential fats in cell membranes and red blood cells. Reduces risk of cancer, heart disease, and other age-associated diseases.	Nuts, seeds, avocado and salad dressings.

Source: The American Dietetic Association and WebMD.

Because of high productivity per unit area compared to cereal crops, fruits and vegetables provide much better economic returns when properly managed besides serving as a good source of foreign exchange earnings. Cultivation of fruits and vegetables provides maximum potential for generating employment in the countryside.

1.2 Area and Production

Of the available 194 million hectares of land for agriculture (total cropped area), the area under fruits and vegetables is about 8 per cent. The diverse agroclimatic zones available in our country make it possible to grow almost all kinds of fruits and vegetables in India (be it tropical, sub-tropical or temperate).

India is the second largest producer of fruits in the world (81.28 million tonnes) with a production share of 12.6 per cent as per the Indian Horticulture Database 2013 published by the National Horticulture Board. During the same period banana recorded the highest production of 26.50 million tonnes, followed by mango (18.00 million tonnes), citrus (10.09 million tonnes), papaya (5.38 million tonnes), guava (3.20 million tonnes), grape (2.48 million tonnes), apple (1.91 million tonnes), pineapple (1.57 million tonnes) and sapota (1.50 million tonnes). Among the vegetables potato occupied the first rank with production of 45.34 million tonnes followed by tomato (18.22 million tonnes) and onion (16.8 million tonnes). India produced 162.18 million tonnes of vegetables during 2012-13 which is 14.0 per cent of the global production. The productivity of grape (21.1 t/ha), pea (9.5 t/ha) and cassava (35.0 t/ha) in India has been reported to be the highest in the world. India is considered as "the land of spices" for being the largest producer (57.44 lakh tonnes/annum) and exporter of spices in the world. In a short span of time, India has also made significant advancement in production of cut flowers and ornamental plants. According to NHB publication, 2013 India produced 1.73 million tonnes of loose flowers and 76.73 million tonnes of cut flowers. Production share of major fruits and vegetables in India is presented in Figure 1 and Figure 2.

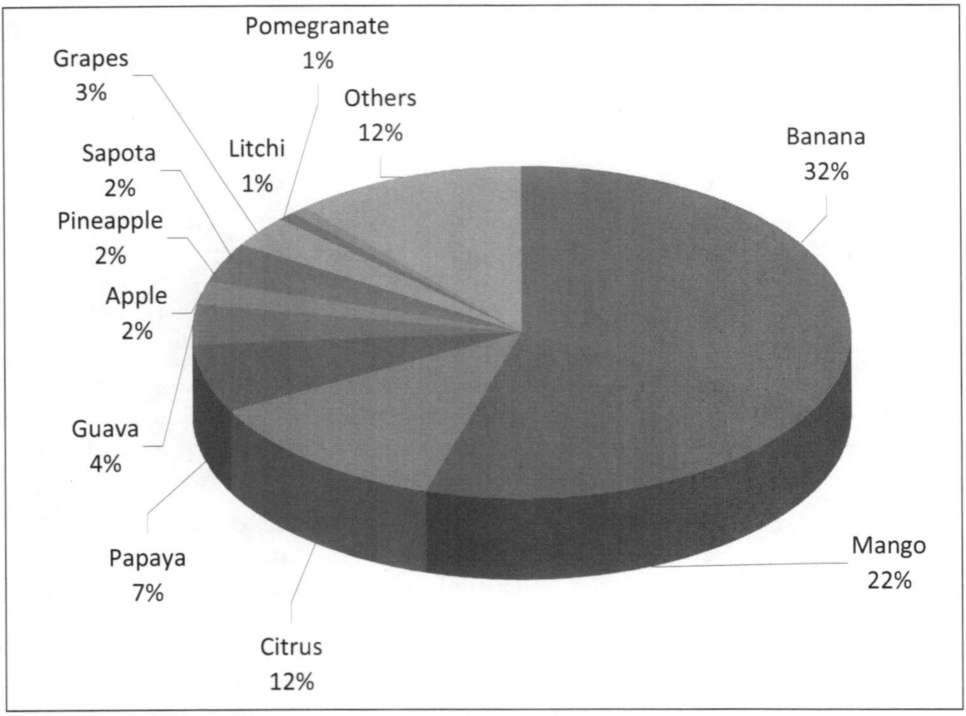

***Source*: Indian Horticulture Database, NHB, 2013**

Figure 1: Production Share of Major Fruits in India (2012-13)

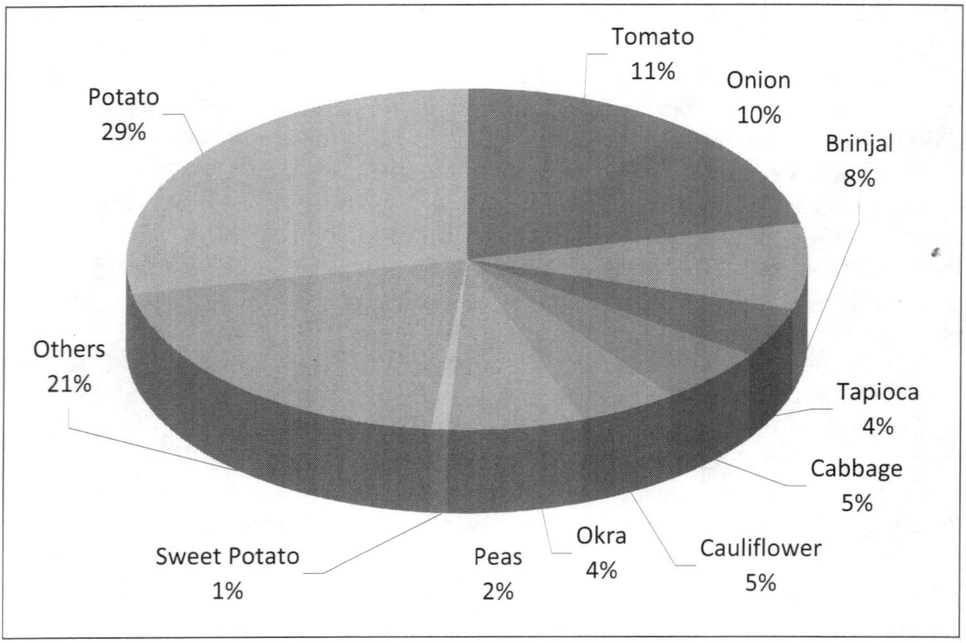

***Source*: Indian Horticulture Database, NHB, 2013**

Figure 2: Production Share of Major Vegetables in India (2012-13)

1.3 Postharvest Losses

Although India has touched new heights in production of fruits and vegetables comparable to Brazil and China, the production is still insufficient to meet the basic nutritional requirements of the growing population. Part of this problem could be overcome by salvaging a fair proportion of huge postharvest losses of these perishables which are equivalent to total annual production of these commodities by United Kingdom. These losses occur at several stages in the postharvest management chain *viz.* harvesting, grading, packaging, transport, storage and marketing. Hence, it is better to pay more attention in managing postharvest losses besides increasing production and productivity of horticultural crops.

In India the estimated postharvest losses of fruit and vegetables have been reported to be within the range of 6-36 per cent depending upon the commodity whereas these range between 16-50 per cent in different countries of Asia-Pacific region. Such losses in developing countries have been reported to be around 32 per cent. Postharvest losses can be divided into two categories *e.g.*, quantitative and qualitative. Qualitative losses include deterioration in texture, flavour, aroma, volatile compounds, nutritional value and cosmetic look. However, quantitative losses are valued at several thousand crores of rupees. It has been estimated that every one percent reduction in postharvest loss in India can save approximately 5 million tonnes of fruits and vegetables.

When we aim to reduce such quantitative and qualitative postharvest losses of fruits and vegetables, we must consider both the biological and environmental factors involved in the deterioration of the produce after harvest and before reaching consumers. Basically postharvest management attempts to slow down the undesirable changes that the commodity may undergo in its postharvest life.

1.4 Factors Affecting Postharvest Life

As almost all edible plant parts are living tissue, there are irreversible postharvest physiological changes leading to senescence/cell death. However management of postharvest processes depending on the plant part consumed demands different postharvest treatments/storage requirements. Additionally all fresh horticultural commodities being high in water content face desiccation causing wilting/shriveling and corresponding postharvest loss. Like all other living tissues, horticultural produce also falls prey to pests and diseases, both before and after harvesting.

Biotic factors which are responsible for postharvest degradation could be mainly identified as: stage of harvest maturity, rates of respiration, transpiration, and ethylene production, vulnerability to ethylene, changes in chemical composition, colour and flavour, tendency of sprouting, physiological reaction towards low temperature storage conditions besides pathological infections.

Respiratory breakdown leads to both quantitative and qualitative losses resulting in weight loss and food value. Moreover, senescence is also accelerated leading to shortening of storage life. Thus crops with low rate of respiration like nuts, apple, onion, garlic, potato, etc. degenerate slowly while others namely, broccoli, green onion, cut flowers, etc. are highly perishable due to very high respiratory rate. Such direct relationship can also be established between rate of ethylene production and ripening of climacteric fruits *e.g.* apple, banana, guava, tomato and sapota. Non-climacteric fruits *e.g.* citrus, grape, litchi, pineapple, pomegranate, strawberry, sweet pepper, etc. liberate ethylene at a very low rate. Therefore, these two types of fruits require different approach to postharvest management. Change in colour during postharvest life could be a desired trait in fruits but not in case of vegetables.

Temperature, relative humidity, gaseous combination of the storage environment, light, type of packaging material, mode of transport, etc. are some other important abiotic factors influencing postharvest life of horticultural produce. Bruising due to mechanical injury is a great cause of concern as it allows microbes to get an entry in the fleshy part. To minimise losses caused by pathogens maintenance of hygiene always remains a priority during postharvest operations.

Appropriate postharvest management is essential not only for controlling various physiological processes *viz.* respiration, transpiration and other metabolic activities but also for quality assurance, prolonging market life, checking market gluts and for increasing financial gains. Postharvest behavior of perishable horticultural crops depends further on species, cultivar, harvest maturity, postharvest treatments and storage conditions. Critical gaps observed in postharvest management in India are poor farm practices, improper harvesting, improper ripening practices, poor postharvest handling, poor infrastructure and logistic facility and ill equipped marketing system.

1.5 Research and Development Infrastructure

Realizing the importance of postharvest management practices in managing Postharvest quality and reducing postharvest losses, a beginning was made by Indian Council of Agricultural Research (ICAR), New Delhi during fourth five-year plan. The All India Coordinated Fruit and Vegetables Improvement Projects were started in 1970-71 by the Indian Council of Agricultural Research. The broad objectives of such projects were development of fruit and vegetable cultivars for higher yields, identification of varieties for disease resistance, better shelf life and to some extent suitability for processing. A large number of research centres were established all over the country under the above project to study scientific methods of reducing wastage during the process of handling, transport, storage and processing for ensuring full utilization of horticultural produce.

A study group under the chairmanship of Dr. M.S. Swaminathan, the then Director General, ICAR recommended setting up of a coordinated research programme on postharvest technology of fruits, vegetables and flowers in the country. The programme came into operation during August, 1978 through PL-480 grant made available under the Agricultural Special Foreign Currency Programme of the United States Department of Agriculture with Dr. J.C. Anand as the first Project Coordinator. The primary objective of the project was to reduce postharvest losses of fruits and vegetables during harvesting, farm handling, storage, marketing, transportation and processing. A Coordinated Project was started with four centers located at Indian Agricultural Research Institute (IARI), New Delhi, Indian Institute of Horticulture Research (IIHR), Bangalore, Central Institute of Horticulture for Northern Plains (CIHNP), currently known as Central Institute of Subtropical Horticulture (CISH), Lucknow and Dr. Y.S. Parmar University of Horticulture and Forestry (YSPUHF), Solan. Considering importance and impact of this project on Indian agri-economy, several more centres were later added in different states during subsequent plans.

During the 7th plan another collaborative project, the Indo-USAID sub-project on postharvest technology of fruits and vegetables, was approved under the Indo-US sub-commission on Agriculture which remained operational at four centres *viz.* IARI, New Delhi, IIHR, Bangalore, CIHNP, Lucknow and NRCC, Nagpur till 1991. In this project, seven economically important horticultural crops *viz.* mango, banana, citrus, guava, potato, onion and tomato were identified. The major objective of the coordinated projects during the period from 1971 to 1991 was to minimize postharvest loss. Subsequently research on postharvest management was started at various State Agricultural Universities in the country. A Central Institute of Postharvest Engineering and Technology was also setup by ICAR at Ludhiana with a regional station at Abohar to intensify research on this aspect. Several other old and new Institutes namely CFTRI, Mysore, Regional Research Laboratories (CSIR), National Agril. Food Biotechnology Institute, Mohali under (DBT), Bhabha Atomic Research Centre, Mumbai, Indian Institute of Crop Processing Technology, Thanjavur and National Institute of Food Technology Entrepreneurship and Management, Sonipat, Haryana, Defence Food Research Laboratory, Mysore are now involved in R&D Programmes on various agri/hort crops.

Indian agriculture scenario was becoming brighter during this period, farmers were reaping the benefits of Green Revolution and suddenly the country had more agricultural produce to manage. To handle the surplus production, stakeholders realized need of investment in storage infrastructures, and postharvest processing plants. Ministry of Food Processing Industries was set up in 1988. To promote export and to deal with related policy matters Agricultural and Processed Food Products Export Development Authority (APEDA) was established by the Government of India under the Ministry of Commerce through an Act passed by the Parliament in December, 1985.

To take care of the horticulture sector a National Horticulture Board (NHB) was set up by the Government of India in 1984 as an autonomous society. Though NHB started promoting commercial horticulture in selected agro-climatic zones, it soon integrated the idea of development of postharvest management infrastructure with area expansion projects. Development of energy efficient cold infrastructure for fresh horticultural produce is also a priority research area identified by NHB.

1.6 Managing Postharvest Losses

According to FAO, nearly one-third of food produced for human consumption (approximately 1.3 billion tonnes per year) is either lost or wasted globally along the food chain from production to consumption. A High Level Panel of Experts on Food Security and Nutrition (HLPE) was created in 2010 to provide UN's Committee on World Food Security (CFS) with evidence based and policy oriented analysis about the causes of food losses and waste (FLW) and to suggest actions to reduce them in order to improve food and nutrition security and the sustainability of food systems. The aim of the HLPE report, given the diversity of contexts, was to help all concerned actors to reduce food losses and waste by identifying the causes and potential solutions that may be implemented, alone or in a coordinated way, by the relevant actors in the food system, including the public and private sectors, civil society, individual producers, wholesalers, retailers and consumers. Successful reduction of food losses and waste will save resources and has the potential to improve food security and nutrition, goals shared with the Zero Hunger Challenge and the post-2015 sustainable development agend.

HLPE defines food losses and waste (FLW) as "a decrease, at all stages of the food chain from harvest to consumption, in mass, of food that was originally intended for human consumption, regardless of the cause". FLW represents at the same time a waste of resources, as well as an environmental issue by itself, as for example food-related waste, as part of urban total waste, that has a significant greenhouse gas (GHG) footprint. FLW also tends to become a symbol of inefficiency, unfairness and unsustainability of food systems with economic, environmental and social concerns. Evidently HLPE report mainly discussed the postharvest loss of food. Causes of FLW were discussed in the report under three headings: micro, meso and macro and these causes are often interrelated.

1.6.1 Micro Level

Micro-level causes can be found all along the food chain, and are the direct, immediate reasons for FLW taking place at a certain point of the chain, resulting from actions (or non-action) of individuals potentially leading to FLW. The major causes are:

i) Poor harvest scheduling and timing; rough and careless handling of the produce.

ii) Inadequate or lack of storage conditions and poor temperature management all along the food chain.

iii) Transport can be a major cause of FLW by introducing a time span between production and consumption and by bringing additional risks of mechanical and heat injury. Time spent because of transport can also lead to decrease of nutritional value.

iv) Conditions within the retail outlet (temperature, relative humidity, lighting, gas composition, etc.) and handling practices have an effect on quality, shelf-life and acceptability of the product.

Micro-level solutions at harvest and postharvest stages involve improved practices, adoption of technical innovations, investments, or a combination of these. When appropriately applied, good agricultural practices at the primary stage of production as well as good manufacturing practices and good hygienic practices during food processing can protect food from contamination or damage. A key intervention all along food chains is to improve storage conditions. Various solutions have been already successfully implemented in many places.

1.6.2 Meso and Macro Levels

Very often causes of FLW are found at "higher" meso- and macro-levels. These include:

i) At meso-level, the lack of equipment and/or of good practices, inadequate organization, coordination and communication between food chain actors *e.g.* transformation that renders the product useless at a later stage of the chain, etc., inadequate infrastructure, maladapted economic conditions along the food chain (product unmarketable, etc.)

ii) More macro-level, systemic causes include the absence of a good, enabling environment to support coordination between actors, investment and improvement of practices.

There is no doubt that preharvest conditions and actions in the field can indirectly lead to losses at later stages in the chain, as production and agronomic practices very often influence quality at harvest, suitability for transport and shipping, storage stability and shelf-life after harvest. Nature of storage losses of some commercially important fruits is presented in Table 2.

Table 2: Nature of Storage Losses in some Commercially Important Fruits

Losses	*Crops*
Chilling Injury	Banana, citrus, guava, mango, papaya, pomegranate, sapota
Rind (also stem end rind) breakdown	Citrus
Loss of colour	Litchi
Desiccation and Decay	Litchi, pomegranate
Internal breakdown	Pineapple
Crown dehydration	Pomegranate

2.0 Postharvest Management

There are several method and approaches for reducing losses of horticultural produce. These could be broadly categorized under the following heads:

2.1 Maturity and Harvesting

The issues related to maturity of horticultural produce begin with standardization of optimum maturity indices that ensure proper size, color, good edible quality and better shelf life. Fruits and Vegetables are eaten at various stages of development *e.g.* seed (fresh and sprouted), leaves, flowers, whole fruits and nuts. Delay in harvesting and over maturity reduces the storage quality. A lot of research has been carried out on various types of maturity indices; however several maturity standards or guidelines prepared by Bureau of Indian Standards (BIS) are not adopted due to lack of awareness. Based on respiratory pattern, the fruits are classified into two major groups *viz.* climacteric fruits *e.g.* apple, banana, guava, tomato, sapota and non-climacteric fruits *e.g.* citrus, grapes, litchi, sweet pepper, pineapple, pomegranate, strawberry etc. Therefore, harvest maturity of these two distinct groups of fruits differs and must be taken into account for judicious postharvest management practices. A wide variety of maturity indices have been used in different crops *viz.* formation of abscission layer in melon, heat units in grape and sweet peas; leaf changes in root, bulb and tuber crops; number of days from flowering in banana, papaya, jackfruit; oil content in avocado; skin color in apple, tomato, peach, chilli, pepper; specific gravity in mango and TSS/acid ratio in grape. However, there is an urgent need for evolving commercially feasible maturity indices for all major crops.

Harvesting of horticultural crops plays a very important role on their postharvest quality. Improper harvesting makes the commodity unattractive. Harvest injuries serve entry points for microorganisms that result in early rotting. A number of fruit harvesters have been developed by various institutes *viz.* Dapoli Mango harvester, IIHR mango harvester, IARI mango harvester, CISH mango harvester, etc. which are commercially used in many places. Some of the physical and mechanical methods used for harvesting of major fruits crops of India are: use of sharp sickle in banana, scissor in grape; shaking of limbs in case of ber, pulling the fruit with a jerk in case of pineapple and use of clippers in litchi, kinnow etc.

2.2 Physico-chemical Treatments

Different pre-treatments before harvest or at postharvest stages have been standardized to keep the produce fresh for long time or to enhance their shelf life and retain their quality. For example, potatoes and onions are sprayed with maleic hydrazide (2000-2500 ppm) 2-weeks before harvest or treated with gamma irradiation (0.15 kGy) after harvest in order to inhibit sprouting during storage. This technology is now commonly used in commercial storage of potatoes and onions. Irradiation treatment has been quite promising and an effective technology which is currently available to overcome quarantine restrictions in international trade. Pretreatment of apple before storage with ethanol vapours is quite common. Vapour heat treatment (VHT) and hot water treatment (HWT) or hot air treatment are very simple and have become popular pre-treatments before storage or transportation of fruits, especially mango and citrus.

2.3 Precooling

The first and foremost important practice of handling after harvest is pre-cooling. The primary objective of pre-cooling is to remove the field heat. Pre-cooling is the process of "rapid" removal of field heat and is usually practiced for fresh fruits, vegetables and flowers immediately after harvest but before shipment or storage or processing. The latest developments in the area of pre-cooling include air handling system for forced air cooling, use of ozonated water in hydrocooling and for washing of fruits and vegetables. The primary advantages of pre-cooling are: (a) removal of field heat/respiratory heat, (b) inhibition of the growth of decay causing organisms, (c) restriction of the enzyme activity, (d) reduction of water loss, (e) reduction in rate of respiration and C_2H_4 liberation and (f) rapid wound healing. The production and action of ethylene from harvested fruits, vegetables and flowers are temperature dependent. Harvested produce kept at 25 °C with 30 per cent RH shows a tendency of 36 times more water loss as compared to that stored at 0 °C with 90 per cent RH. Hence, pre-cooling serves as an essential practice in any successful cool chain management of horticultural produce. The speed of cooling depends upon the following factors: (i) accessibility of product to the refrigerating medium. (ii) difference in temperature between the product and refrigerating medium. (iii) velocity of refrigerating medium and (iv) type of cooling medium. In order to predict the end point of pre-cooling it is essential to know the "half-cooling time". Half-cooling time is the time required to reduce the temperature difference between commodities and the coolant by one-half. This is independent of the initial temperature of the commodity. The shape, density, and surface area/volume ratio of the fruit or vegetable and the type of packaging are important factors of determining the appropriate type of cooling medium, *i.e.* air, water or ice. Some products do not tolerate exposure to water. In general, the greater the amount of packaging, the slower the cooling rate. If the packaging is exposed to water, packaging in paper-based material is not recommended unless it is waxed or coated with polymer as lamination.

The two most important factors in pre-cooling are temperature and time, *i.e.* fruit or vegetable must be cooled in the shortest time possible. Product cooling follows a logarithmic function, with initial rapid cooling followed by a slower rate. Since it is difficult to remove all the field heat, pre-cooling to 7/8th of the recommended storage temperature is advisable. The product can then be placed in storage where the remaining 1/8th of the heat can be gradually removed with less energy cost. If several products are being handled, then a system must be selected that is compatible to all of them. Highly perishable products, which have a high respiratory rate need fast cooling and require high refrigeration capacity and fast pre-cooling methods. There are basically five methods used for horticultural commodities. These are (i) Room/air cooling. (ii) Water/hydro- cooling (iii) Forced air-cooling (iv) Vacuum cooling and (v) Package icing.

2.3.1 Room Cooling

It is the practice of storing bulk or packaged produce in a refrigerated room for an indefinite period. By the processes of conduction and convection, the heat is gradually removed from the produce. Because the outside of the containers cool more rapidly than the inside, the cooling is uneven and slow. The cooling rate of room cooling may be sufficient for items such as potatoes, cabbage, or root crops, however, it has been proven unsatisfactory for highly perishable items such as strawberries, blueberries, and snap beans.

2.3.2 Hydro-Cooling

Cooling freshly harvested produce by flooding, immersing, or spraying with large quantities of cold water is done in hydrocooling. Compared to other cooling methods, hydrocooling is fast and generally thorough. Hydrocooling is limited to those produce items that will tolerate liquid water (except onion and garlic).

2.3.3 Package Icing

It is cooling fresh produce by the addition of crushed ice or a slurry of crushed ice and water over the top of a load or to each individual package. This method is limited to the produce that is not harmed by contact with ice. This is commercially used in many cut flowers.

2.3.4 Forced-Air Cooling

Forced-air cooling is a commonly used cooling method that utilizes specially constructed portable fans to draw chilled air horizontally through pallets or stacks of packaged produce. A properly designed forced-air cooling system is fast, energy efficient and relatively inexpensive. It may be utilized on most types of produce.

2.3.5 Vacuum Cooling

When warm produce is placed inside a closed container and reduced to a partial vacuum, a small portion of the water in the produce evaporates, causing a cooling effect. Vacuum cooling is fast and may be utilized very effectively on packaged produce. It is generally more effective on items with a large surface area to volume ratio such as lettuce and various greens.

2.4 Sorting and Grading

Sorting and grading of fruits and vegetables based on size, shape, color, texture, maturity, flavor and aroma is essential for better marketing. Several fruit and vegetable graders have been developed at IARI – New Delhi, IIHR – Bangalore, PKDV – Akola, JNKVV – Jabalpur and GAU – Junagarh, Gujarat. Apart from this several machineries are available for grading like ultrasonic devices, colour sorters, machine vision technology etc.

2.5 Ripening

Artificial fruit ripening is another important issue which needs immediate attention particularly with respect to food safety. Rule 44-AA of the Prevention of Food Adulteration (PFA) Rules, 1955 prohibits the use of acetylene gas commonly known as carbide gas produced from calcium carbide for artificial ripening of fruits. It states that "*No person shall sell or offer or expose for sale or have in his premises for the purpose of sale under any description, fruits which have been artificially ripened by use of acetylene gas, commonly known as carbide gas*". Calcium Carbide is used in some countries as source of acetylene gas, which is an artificial ripening agent. However, acetylene is not nearly as effective for ripening as is ethylene, and acetylene is not a natural plant hormone like ethylene. Also, calcium carbide may contain traces of arsenic and phosphorus, both of which are highly toxic to humans. The mango cultivars ripened with calcium carbide had significantly low carotenoid content and less firmness when compared to naturally ripened fruits (Table 3).

Table 3: Effect of Ripening with Calcium Carbide on Quality of Mango Fruits

Cultivar	*Treatments*	*Sugar/Acid Ratio*	*Total Carotenoids µg/100g pulp*	*Firmness (lbs)*
Alphonso	Natural ripening	82.70	16,450	8.8
	Calcium carbide	43..20	12,800	4.4
Baneshan	Natural ripening	83.75	3,540	7.9
	Calcium carbide	35.90	2300	6.2
Langra	Natural ripening	118.90	4,687	12.8
	Calcium carbide	36.50	2,375	7.0
Pairi	Natural ripening	51.20	11,250	11.6
	Calcium carbide	37.00	8,450	4.8
Totapuri	Natural ripening	87.50	3,500	9.7
	Calcium carbide	45.00	2,250	6.6

Source: Shantha Krishnamurthy and Rao, 1981.

The following acute (short-term) health effects may occur immediately or shortly after exposure to calcium carbide

i) Contact with calcium carbide can irritate the skin causing a rash, redness and burning feeling

ii) Contact can also severely irritate and burn the eyes with possible permanent damage (corneal opacities)

iii) Exposure can irritate the mouth, nose and throat.

iv) Inhaling calcium carbide can irritate the lungs causing cough and/or shortness of breath. Higher exposures may cause a build-up of fluid in the lungs (pulmonary edema), a medical emergency, with severe shortness of breath.

v) It is flammable and reactive and thus a dangerous fire and explosion hazard.

vi) When exposed to water or moisture it forms flammable acetylene gas.

The following chronic (long-term) health effects may also occur some time after exposure to calcium carbide and can last for months or years.

i) Cancer hazard

ii) Reproductive hazard

Use of calcium carbide though banned in India has ben reported to be in use in many crops in different countries. It is however, not clear whether the use of calcium carbide is legal in these countries or not. However, several countries use calcium carbide for ripening of banana, mango, citrus, tomato, plums and peaches (Table 4).

Table 4: Countries where the Fruits Continue to be Ripened Using Calcium Carbide

Fruits	*Countries*
Banana	Australia, Egypt, India, Philippines, South Africa, Sudan, Taiwan, U.S.A, Yemen
Citrus	Australia, Philippines, South Africa
Mango	Brazil, Costa Rica, India, Malaysia, Philippines, Senegal, South Africa
Peach	South Africa
Plum	South Africa
Tomato	Australia, Morocco, Philippines, U.S.A

Source: Rahman *et al*., 2008.

2.6 Packaging

Packaging of fresh fruits and vegetables is one of the most important steps in the long and complicated journey from grower to consumer. Bags, crates, hampers, baskets, cartons, bulk bins, and palletized containers are convenient for handling, transporting, and marketing fresh produce. Although the industry generally agrees that container standardization is one way to reduce cost, the trend in recent years has moved toward a wider range of package sizes to accommodate the diverse needs of wholesalers, consumers, food service buyers, and processing operations. Packing and packaging materials contribute a significant cost to the produce. It is therefore important that packers, shippers, buyers, and consumers have a clear understanding of the wide range of packaging options available. A significant percentage of buyer and consumer complaints may be traced to container failure because of poor design or inappropriate selection and use. A properly designed produce container should contain, protect, and identify the produce, satisfying everyone from grower to consumer.

The most important considerations in packaging are:

2.6.1 Recyclability

Many developed countries and export markets have waste disposal restrictions for packaging materials. In the near future, almost all produce packaging may have to be recyclable or biodegradable, or both.

2.6.2 Sales Appeal

Packaging plays a vital role in attracting the consumers. High quality graphics are increasingly becoming popular to boost sales appeal. Multi-color printing, distinctive lettering, and logos are now common. Modern produce packaging can be custom engineered for each commodity to extend shelf life and reduce waste.

2.6.3 Containment

The package must enclose the produce in convenient units for handling and distribution. Optimization of space inside the package to accommodate produce is very important. Small produce items that are spherical or oblong (such as potatoes, onions, and apples) may be packaged efficiently utilizing a variety of different package shapes and sizes. However, many produce items such as asparagus, berries, or soft fruits may require containers specially designed for that item. Packages of produce commonly handled by hand are usually limited to 25 kg. Bulk packages that are moved by fork lifts may weigh as much as 600 kg.

2.6.4 Protection from Mechanical Damage

The package must protect the produce from mechanical damage and poor environmental conditions during handling and distribution. To buyers, torn, dented, or collapsed produce packages usually indicate lack of care in handling the contents. Produce containers must be sturdy enough to resist damage during handling, storage, and transportation to market. Because almost all produce packages are palletized, produce containers should have sufficient stacking strength to resist crushing in a low temperature, high humidity environment. Although the cost of packaging materials has escalated sharply in recent years, poor quality, lightweight containers that are easily damaged by handling or moisture are no longer tolerated by packers or buyers. Produce destined for export markets requires containers to be extra sturdy. Air-freighted produce may require special packing, with respect to package sizes, and insulation. Marketers who export fresh produce should consult with freight companies about any special packaging requirements. APEDA of Government of India may also be able to provide specific packaging information.

Damage resulting from poor environmental control during handling and transit is one of the leading causes of rejected produce and low buyer and consumer satisfaction. Each fresh fruit and vegetable commodity has its own requirements for temperature, humidity, and environmental gas composition. Containers should help to maintain an optimum environment for the longest shelf life of produce. This may include special materials to slow down the loss of water from the produce, insulation materials to keep out the heat, or engineered plastic liners that maintain a favorable mix of oxygen and carbon dioxide.

2.6.5 Information about Produce

The package must identify and provide useful information about the produce. It is customary (and may be required in some cases) to provide information such as the produce name, brand, size, grade, variety, net weight, count, grower, shipper, and country of origin. It is also becoming common to give nutritional information, recipes, and other useful information on the packages directed specifically at the consumer. In consumer marketing, package appearance has also become an important part of point of sale displays. Universal Product Codes (UPC or bar codes) may be included as part of labeling. The UPCs used in the food industry consist of a ten-digit machine readable code. The first five digits are a number assigned to the specific producer (packer or shipper) and the second five digits represent specific product information such as type of produce and size of package. Although no price information is included, UPCs are used more and more by packers, shippers, buyers, and retailers as a fast and convenient method of inventory control and cost accounting. Efficient use of UPCs requires coordination with everyone who handles the package.

2.7 Packaging Materials

2.7.1 Wooden Pallets

Pallets literally form the base on which most fresh produce is delivered to the consumer. Pallets were first used during World War II as an efficient way to move goods. Standard size pallets make efficient use of truck and van space and can accommodate heavier loads and more stress than lighter single-use pallets. Substantial wooden pallet bins of milled timber or plywood are primarily used to move produce from the field or orchard to the packing house. Although the height may vary, the length and width is generally the same as a standard pallet (120 cm by 100 cm). More efficient double-wide pallet bins (120 cm by 200 cm) are becoming more common in some produce operations. Most pallet bins are locally made; therefore it is very important that they be consistent from lot to lot in materials, construction and especially size. For example small differences in overall dimensions of pallet bin can add up to big problems when several hundred are stacked together for cooling, ventilation, or storage. It is also important that stress points be adequately reinforced. When properly protected from the weather, pallets bins may have a useful life of 10 years or more. Wooden crates, once extensively used for apple, mango, stone fruit and potatoes have been almost totally replaced by other types of containers.

2.7.2 Corrugated Fiber Boxes (CFB)

Corrugated fiber board (often called cardboard or pasteboard) is manufactured in many different styles and forms. Due to its relativity low cost and versatility, it is the dominant produce container material. Corrugated fiberboard is made from three or more layers of paperboard. Double-faced corrugated fiberboard is the predominant one. It is produced by sandwiching a layer of corrugated paperboard between an inner and outer liner (facing) of paper-board. Unless the CFB boxes are specially treated, moisture absorbed from the surrounding air and the contents can reduce the strength of the container by as much as 75 percent. New anti-moisture coatings with plastic substantially reduce the effects of moisture.

There are numerous types of corrugated fiberboard containers. The two types most used in the industry are the one piece, regular slotted container (RSC) and the two piece, full telescoping container (FTC). The RSC is the most popular because it is simple and economical. However, it has relatively low stacking strength and therefore must be used with produce, such as potatoes, that can carry some of the stacking load. The FTC (with one container inside another), is used when greater stacking strength and resistance to bulging is required. A third type of container is the Bliss box, which is constructed from three separate pieces of corrugated fiberboard. The Bliss box is used when maximum stacking strength is required. The bottoms and tops of all three types of containers may be closed by glue, staples, or interlocking slots.

2.7.3 Shrink Wrapping

One of the newest trends in produce packaging is the shrink wrapping of individual produce items. Shrink wrapping has been used successfully to package, apple, onion, sweet corn, cucumber and a variety of tropical fruits and vegetables. Shrink wrapping with an engineered plastic wrap can reduce shrinkage, protect the produce from disease, reduce mechanical damage and provide a good surface for stick-on labels.

3.0 Spoilage of Produce

Different types of micro-organisms can cause spoilage in storage houses. However, bacteria and fungi are most common. Bacteria are generally propagated by direct contact between perishable goods or with contaminated surfaces, or else with the water used during treatment before or after storage. Some of these bacteria can be pathogenic for human as well, such as *Listeria monocytogenes,* sometimes found on certain vegetables. Vegetables are mostly spoiled by bacteria due to their high pH (4.5-7.0) whereas many fruits particularly the acidic fruits (pH < 4.5) inhibit the growth of bacteria but encourage fungal spoilage. A relative humidity of less than 94 to 95 per cent is considered low enough for growth of bacteria. Majority of the soft rot causing bacteria have their optimum temperature for 30 °C. The following are the major bacterial species that infect the fruits and vegetables (Table 5).

Table 5: Major Bacterial Diseases of Horticultural Produce

Commodity	*Causal Organism*
Apple and pear	*Pseudomonas* spp.
Cabbage, lettuce	*Pseudomonas cicorii*
Cauliflower, cabbage, tomato, bean, radish	*Xanthomonas* spp.
Mango, plum, cherry, peach	*Xanthomonas campestris*
Onion	*Pseudomonas cepacia*
Pineapple, papaya, guava	*Ewrinia* spp.
Potato	*Erwinia caratovora*
	Bacillus subtilis
	Clostridium sp.

3.1 Postharvest Diseases

3.1.1 Bacterial Diseases

Chlorine compounds added to water as chorine gas or as hypochlorite solution could be used to kill bacteria rapidly. In commercial practice 2 per cent sodium hypochlorite or 50 to 100 mg of chlorine or 1-1.5 g of chlorine dioixde per litre of water are frequently used to control the bacteria. Low temperature storage is considered to be the best method for controlling spoilage due to bacterial diseases.

3.1.2 Fungal Diseases

Several fungal rots cause severe postharvest losses to fruits and vegetables by latent or quiescent infection. Some of the examples of quiescent infections are mentioned below (Table 6).

Table 6: Major Fungal Organisms Affecting Fruits in Quiescent Infection

Commodity	*Causal Organism*
iApple	*Gloeosporium perennans, Diplodia perniciosa, Nectria galligena*
Citrus	*Colletotrichum gloeosporioides*
Mango	*Colletotrichum gloeosporioides*
Strawberry	*Botrytis cinerea*

Control of quiescent infection in fruits can be achieved by

i) Strict sanitation in pack house to reduce conidial rot
ii) Avoidance of injury
iii) Postharvest dips with permitted systemic fungicides *viz.* Prochloraj carbendazim etc.
iv) Pre-cooling or cool chain management

Two major diseases of mango observed during postharvest stage are: a) anthracnose: caused by *Glomerella cingulata* (conidial stage *Colletotrichum gloeosporioides*) and the stem end rot of mango associated with *Botryodiplodia theobromae* or *Aspergillus niger*. Among the other pathogens *Penicillium* spp. is responsible for causing diseases to many fruits and vegetables *e.g.* apples, citrus etc. Finger rot of banana caused by *Botryodiplodia theobromae.*

Major fungal organisms mostly associated with postharvest diseases of vegetables are:

i) *Fusarium* spp. causing rots in round melon, bean, cowpea, chilli, tomato, cauliflower, potato, cucumber, pointed gourd and pumpkin.
ii) *Geotrichum* spp. which causes spoilage in tomato, pointed gourd, watermelon and muskmelon attack okra and cucurbits.

The following are the time temperature combinations for control of the fungal diseases in fruit and vegetables (Table 7).

Table 7: Time-Temperature Combinations for Control of Fungal Diseases in Fruit Crops

Organism	*Temperature (°C)*	*Time/Duration (min)*
Bortyodiolodia, Gloeosporium, Colletotrichum, Rhizopus	43-49 (Papaya)	20
Colletotrichum, Diplodia, Aspergillus, Bortyodiolodia, Rhizopus	47-55 (Mango)	10-20
Monilia, Rhizopus	49-84 (Peach)	1.5-3.5
Alternaria, Geotrichum	48-57 (Tomato)	5-10

In the recent times consumers are very much concerned about residual toxicity of pesticides on fruits and vegetables. Several bio-control formulations available in the market which could be successfully used in postharvest management of horticultural produce are listed in Table 8.

Table 8: Biological Control of Postharvest Diseases

Crops	*Name of Disease*	*Biocontrol Agent*
Apricot, avocado, cherry, citrus, litchi, Strawberry	Brown rot, Grey mold, Green Mold, Stem end rot, Alternaria rot	*Bacillus subtilis*
Mango	Anthracnose, steam end rot	*Bacillus lichenifomis*
Citrus, mango, strawberry	Green mold, Stem-end rot, Gray mold	*Tricoderma viride*
Apple, citrus, peach	Blue mold, Green and blue mold and Gray mold, Brown rot	*Pseudomonas syringae*
Banana, citrus, papaya	Penicillium rots and Crown rot, Anthracnose	*Candida oleophila*
Citrus, peach	Green and blue mold, Blue mold, Rhizopus rot, Sour rot	*Debaryomyces hansenii*

3.2 Insect Pests

A large number of insect pests adversely affect the postharvest quality of horticulture produce. These include fruit fly, mite damage in citrus, thrip infection in banana, mealy bug in pineapple etc. Adoption of strict regulation is required to eliminate the risk of entry of new pests. Although fumigants such as ethylene dibromide (EDB) and methyl bromide (MB) are often used for disinfestation in many countries these are slowly being phased out due to their carcinogenic effects. Quarantine treatments for fruit fly is mandatory except in European Union and Canada for fruits like mango, papaya, melons and many other fruits and vegetables. Hence there is need for chemical free produce for its acceptability in the international market. Recent developments on irradiation and insecticidal controlled atmosphere are some of the technologies for pest management strategies of fruits and vegetables.

Use of irradiation in postharvest management is limited to inhibition of sprouting in potato, onion, garlic, and ginger, reduction in pathogen inoculum in spices and disinfestation of insects in dry fruits. However, it has also been used as quarantine treatment for export of mango from India to United States. Technologies of application of irradiation on fresh pre-packed foods have been standardized by BARC, Mumbai. Several researches have proved that irradiation helps delaying ripening and senescence and controlling fugal decay. More than 21 horticultural products are now being treated with gamma irradiation about 45 countries in the world use this technology. Studies on wholesomeness, microbial safety, nutritional adequacy, mutagenicity and toxicity have shown that this technology is safe.

Vapour Heat Treatment (VHT) has been developed in Japan in the year 1993 for destruction of fruit flies in green pepper. This technology has now been emphasized to disinfest insects in several kinds of fruits exported from Thailand, Philippines, Taiwan, Hawaii, Fiji, Mexico, Australia, Brazil, and Columbia. This treatment eliminates the use of chemicals for disinfestation of fruits flies. VHT has been widely accepted in most of the countries where imports of fruit are banned due to chemical disinfestation of fruit flies. In India the facilities of VHT or hot water treatment of mango are available and standardization of VHT for different fruits needs to be done.

Several chemical growth regulators are being used for enhancing shelf life and delaying ripening of fruits and vegetables. Some examples of chemicals and edible surface coating materials used in fruits and vegetables for regulation of postharvest behaviour are mentioned in Table 9.

Table 9: Important Edible Surface Coating Materials

Coating Material	*Fruits*
Semperfresh	Apple, Banana, Guava
Brilloshine	Apple, Avocado, Citrus fruits
Citrashine, Nu-coatflo	Citrus fruits
Tal prolong, Applewax	Apple
Nutri-save	Apple (Golden Delicious)
Vapor gard	Mango
Ban-seel	Banana and Plantains
Chitosan	Raspberry, Strawberry

In recent times several new molecules have been used to retain the postharvest quality of fruits. These are Methyl Jasmonate (MeJA) for alleviation of chilling injury, *viz.* loquat, peach, pomegranate, strawberry. Similarly, inducing resistance for postharvest diseases and checking browning in cut fruit salicylic acid (SA) has been found to be useful, in case of banana, cherry, kiwi fruit, mango, and pomegranate. Use of nitric oxide (NO) has been found successful in peach for delay in fruit softening, reduction in chilling injury, microbial infection and extension of postharvest life (Table 10).

4.0 Storage

4.1 Controlled (CA)/Modified Atmosphere Storage (MAS)

Postharvest losses of fruits and vegetables during storage occur primarily due to chilling injury in case of banana, citrus, guava, mango, papaya, pomegranate, sapota: rind breakdown in case of citrus; loss of peel color in litchi, desiccation and decay in litchi and pomegranate; internal breakdown and crown dehydration in pineapple, pomegranate etc. The effect of gas on harvested fruits is known since ancient times. In earlier days fruits were taken to temples for improving ripening. The scientific explanation of this phenomenon could be attributed to the volatiles released on burning of incense containing hydrocarbon gases causing ripening of fruits. Earliest scientifically documented evidence on postharvest physiology was recorded by J.E. Bernard in 1819 in France who noticed that harvested fruits absorb O_2 and liberate CO_2. Atmosphere devoid of O_2 caused no ripening in peach, prunes and apricot for several days but ripening continued when they were placed back in air. A commercial cold storage built in 1856 by B. Nice used ice to maintain a temperature of

Table 10: Use of Chemicals in Postharvest Management

Chemicals	*Crops*	*Effect*
Calcium Chloride	Grape	Improves shelf life
	Pineapple	Better ripening and overall quality
CIPC	Potato	Inhibits sprouting
Ethephon	Grape	Hastens ripening
Ethylene	Tomato	Hastens ripening
GA	Guava, Mandarin and Mango	Delays ripening
Guar gum	Pomegranate	Reduces weight loss
Methyl Jasmonate	Loquat, peach, pomegranate, strawberry	Checks postharvest decay, alleviates chilling injury, delays fruit ripening and improves sensory apeal
MH	Onion, Potato	Checks sprouting
N-Benzyladanine	Leafy vegetables	Improves shelf life
Nitric Oxide (NO)	Peach	Delays fruit softening, chilling injury, microbial infection and extends postharvest life.
Polyamines	Pomegranate	Alleviates chilling injury, reduces respiratory rates and retards colour chang
Salicylic Acid (SA)	Banana, cherry, kiwi, mango, pomegranate	Increases resistance to diseases, delays ripening and inhibits browning in cut fruits
Skin coatings with waxes	Citrus, Mango	Reduces respiratory rate and moisture loss
Smoking	Banana	Enhances ripening

1 °C. After a decade he experimented with modifying the cold store with CO_2 and O_2 gases inside. In 1907 Foulton J. observed the increase in fruit damage by large accumulation of CO_2 in the storage atmosphere. In 1915 R.W.Thatcher on their experiments with apples in sealed boxes containing various levels of gases concluded that CO_2 greatly inhibited ripening. The first scientific evidence on the effect of CO_2 on respiratory rates was established by Kidd and West in 1917. In 1927 they carried out comprehensive studies on the effect of CO_2 and O_2 on storage of fruits and vegetables. In the early 1940 the term 'gas storage' was replaced by Controlled Atmosphere storage.

Atmosphere at ambient conditions comprises of 78.08 per cent N_2, 20.98 per cent O_2 and 0.03 per cent CO_2 under normal conditions. Any deviation from this normal atmosphere composition *e.g.* elevated level of CO_2 reduced level of O_2, N_2 or any other combination is known as '**Modified Atmosphere**'. When this deviated normal atmosphere is precisely kept under control then it is termed as "**Controlled Atmosphere**". This control can be done in package (Controlled Atmosphere Packaging) or in the storage chamber (CA-storage). Similar is the case for Modified Atmosphere Storage (MA-storage) and Modified Atmosphere Packaging (MAP).

Generally, O_2 below 8 per cent and CO_2 above 1 per cent are used in CA-storage. Atmospheric modification is a supplementary practice to temperature management in preserving quality and safety of fresh fruits, vegetables, ornamentals and their products throughout postharvest handling. Essentiality of CA/MA technology should be justified only if:

i) the commodities are having high market value
ii) it significantly enhances storage life
iii) it retains significantly better quality
iv) it fetches better price compared to conventional cool stored produce.

Retardation of ripening, reduction in decay, prevention of specific disorders and maintenance of product texture are some of the potential advantages of CA/MA storage. However, initiation or aggravation of certain physiological disorders *viz.* black heart in potato, brown stain in lettuce, brown heart in apples and pears etc. take place in CA-storage if appropriate gaseous regimes are not maintained. A concentration of less than 2 percent O_2 or more than 5 per cent CO_2 in the storage/package atmosphere results in irregular ripening of banana, tomato and pear fruits. Too low O_2 or too high CO_2 can increase the susceptibility to decay causing organisms. Off-flavour development in fruits, stimulation of sprouting and retardation of periderm formation in potatoes are some of the ill effects of improper CA-condition. On the other hand, a very high concentration of CO_2 (upto 30 per cent) was found to be extremely beneficial in checking the *Botrytis* rot in strawberry fruits.

Controlled atmosphere storage and preservation of packaged horticultural produce is a widely utilized technology for fresh fruits, vegetables and flowers. The utilization of inert gases or vacuum can allow unique applications that control microorganisms as well as maintain product colour and freshness. It should be kept in mind that CA and MA are the supplementary practices to low temperature storage. The quality retention is mainly due to reduction in the respiratory and metabolic activities and check in ethylene liberation by fresh horticultural produce during storage.

4.1.1 Biological Basis

Exposure of fresh horticultural crops to low O_2 and/or elevated CO_2 atmospheres within the range tolerated by each commodity reduces their respiration and ethylene production rates. However, outside this range respiration and ethylene production rates can be stimulated indicating a stress response. This stress can contribute to incidence of physiological disorders and increased susceptibility to decay. Elevated CO_2-induced stresses are additive to and sometimes synergistic with stresses caused by low O_2; physical or chemical injuries; and exposure to temperatures, RH, and/or ethylene concentrations outside the optimum range for the commodity.

The shift from aerobic to anaerobic respiration depends on fruit maturity and ripeness stage (gas diffusion characteristics), temperature, and duration of exposure to stress-inducing concentrations of O_2 and/or CO_2. Up to a point, fruits and vegetables are able to recover from the detrimental effects of low O_2 and/or high CO_2 stresses (fermentative metabolism) and resume normal respiratory metabolism upon transfer to air. Plant tissues have the capacity for recovery from the stresses caused by brief exposure to fungistatic CO_2 (> 10 per cent CO_2). Climacteric fruits are less tolerant and have lower capacity for recovery following exposure to reduced O_2 and/or elevated CO_2 levels than non-climacteric fruits. The speed and extent of recovery depend upon duration and levels of stresses, stage of maturity and metabolically driven cellular repair mechanism.

Elevated-CO_2 atmospheres inhibit activity of ACC synthase (key regulatory enzyme of ethylene biosynthesis), while ACC oxidase activity is stimulated at low CO_2 and inhibited at high CO_2 concentrations and/or low O_2 levels. Thus elevated CO_2 atmospheres inhibit ethylene action. Optimum modified atmospheric compositions retard chlorophyll loss (green color), biosynthesis of carotenoids (yellow and orange colours), anthocyanins (red and blue colors), biosynthesis and oxidation of phenolic compounds (brown color). Controlled atmospheres slow down the activity of cell wall degrading enzymes involved in softening and enzymes involved in lignification leading to toughening of vegetables. Low O_2 and/or high CO_2 atmospheres influence flavor by reducing loss of acidity, starch to sugar conversion, sugar inter-conversions, and biosynthesis of flavor volatiles.

Severe stress CA conditions decrease cytoplasmic pH and ATP levels, and reduce pyruvate dehydrogenase activity while pyruvate decarboxylase, alcohol dehydrogenase, and lactate dehydrogenase are induced or activated. This causes accumulation of acetaldehyde, ethanol, ethyl acetate, and/or lactate, which may be detrimental to the commodities if they are exposed to stress CA conditions beyond their tolerance. Specific responses to CA depend upon cultivar, maturity and ripeness stage, storage temperature and duration, and in some cases, ethylene concentrations.

Nitrogen is an inert component of CA. Replacing N_2 with argon or helium may increase diffusivity of O_2, CO_2 and C_2H_4, but they have no direct effect on plant tissues and are more expensive than N_2 as a CA component. Super-atmospheric levels of O_2 up to about 80 per cent may accelerate ethylene-induced de-greening of non-climacteric commodities and ripening of climacteric fruits, respiration and ethylene production rates, and incidence of some physiological disorders (such as scald on apples and russet spotting on lettuce). At levels above 80 per cent of O_2 some commodities and postharvest pathogens suffer from O_2 toxicity. Use of super-atmospheric O_2 levels in CA will likely be limited to situations in which they reduce the negative effects of fungistatic or elevated CO_2 atmospheres on commodities that are sensitive to CO_2-injury.

4.1.2 Commercial Application

Controlled atmosphere (CA) storage systems are used commercially for the long-term storage of fresh horticultural crops. Commercial use of CA storage is maximum in apples and pears worldwide. Recent research has shown the potential advantages of this method in short-term (a few days) and medium-term (a few weeks) storage of certain types of produce. Optimizing storage conditions requires facilities that allow the temperature and composition of gases in storage rooms to be controlled precisely. Each product reacts in different ways to different concentrations of gases.

There are various types of CA systems used in commercial storage. These include generating nitrogen by separation from compressed air using molecular sieve beds or membrane systems. Various types of CA storage system include ultra low O_2 (1.0 to 1.5 per cent) storage, low ethylene (< 1 $\mu L\ L^{-1}$) CA storage; rapid CA storage (rapid establishment of optimal levels of O_2 and CO_2), and programmed or sequential CA storage (*e.g.*, storage in 1 per cent O_2 for 2 to 6 weeks followed by storage in 2 to 3 per cent O_2 for the remainder of the storage period) etc. Recent reports of short-term control atmosphere exposure (CAE) techniques indicate a great promise of simulated effect of continuous CA–storage system with particular reference to delay in disease development, delaying the senescence and quality assurance of CA-insensitive climacteric fruits. Other developments include use of atmospheric modification during transport and distribution, improved technologies of establishing, monitoring, and maintaining CA, using edible coatings or polymeric films with appropriate gas permeability to create a desired atmospheric composition around the commodity.

4.2 Modified Atmosphere Packaging

MAP may be defined as the enclosure of food products in gas-barrier materials, in which the gaseous environment has been changed in order to inhibit spoilage agents and therefore either maintain a higher quality within a perishable food during its natural life or actually extend the shelf life. Although the terms Controlled Atmosphere (CA) and Modified Atmosphere (MA) are often used interchangeably, a precise difference exists between these two terms.

4.2.1 Relative Tolerance to O_2 and CO_2

The extent of benefits from CA and MA use depend upon the commodity, cultivar, physiological age (maturity stage), initial quality, concentration of O_2 and CO_2, temperature, and duration of exposure to such conditions. Subjecting a cultivar of a given commodity to O_2 levels below and/or CO_2 levels above its tolerance limits at a specific temperature time combinations will result in stress to the living plant tissue, which is manifested as various symptoms. Tables 11 and 12 include classifications of fruits and vegetables according to their relative tolerance to low O_2 or elevated CO_2 concentrations when kept at or near their optimum storage temperature and relative humidity.

Table 11: Classification of Fruits and Vegetables According to their Tolerance to Low O_2 Concentrations

Minimum O_2 Concentration Tolerated (per cent)	*Commodities*
1.0	Some cultivars of apple and pear, broccoli, garlic, mushroom, onion
2.0	Most cultivars of apple, apricot, Brussels sprout, cabbage, cantaloupe, cauliflower, celery, cherry, green bean, kiwifruit, lettuce, nectarine, olive, papaya, peach, pear, pineapple, plum, strawberry
3.0	Artichoke, avocado, cucumber, pepper, persimmon, tomato
5.0	Asparagus, citrus fruits, green pea, potato, sweet potato

Source: Kader *et al.*, 1989.

Table 12: Classification of Fresh Fruits and Vegetables According to their Tolerance to Elevated CO_2 Concentrations

Maximum CO_2 Concentration Tolerated (per cent)	*Commodities*
2	Apple (Golden Delicious), Pear (Asian and European), apricot, grape, olive, tomato, sweet pepper, lettuce, Chinese cabbage, artichoke, sweet potato
5	Most cultivars of Apple, avocado, banana, brussel's sprouts, cabbage, carrot, cauliflower, chilli, egg plant, kiwifruit, mango, nectarine, orange, papaya, pea, peach, plum, radish
10	Asparagus, broccoli, cucumber, , garlic, grape fruit, leek, lemon, lime, okra, onion (dry), onion (green), persimmon, pineapple, potato
15	Blackberry, cantaloupe, fig, kale, mushroom, raspberry, spinach, strawberry.

Source: Kader *et al.,* 1989.

Successful MAP or CA must maintain near optimum O_2 and CO_2 levels to attain the beneficial effects of modified atmosphere without exceeding the limits of tolerance which may increase the risk of physiological disorders and other detrimental effects.

4.3 Creating Modified Atmosphere Conditions

Modified atmosphere can be created either passively by the respiration of commodity also called commodity generated modified atmosphere or active modified atmosphere generation.

4.3.1 Passive Modified Atmosphere

If commodity characteristics are properly matched to film permeability characteristics, an appropriate atmosphere can passively evolve within a sealed package as a result of the consumption of O_2 and the production of CO_2 through respiration. This atmosphere must be established rapidly and without danger of the creation of anoxic conditions on injuriously high level of CO_2.

4.3.2 Active Modified Atmosphere

Because of the limited ability to regulate a passively established atmosphere, an active modified atmosphere can be created by vacuum packaging or replacing the package atmosphere with the desired gas mixture.

a. Vacuum Packaging

Involves placing a product in a film of low O_2 permeability, the removal of air from the package and the application of a hermetic seal. Product compression is unavoidable and it makes vacuum packaging unsuitable for many products.

b. Gas Packaging

Gas packaging overcomes the limitations of vacuum packaging. Oxygen, nitrogen and carbon dioxide are gases generally used which have specific role *viz.* O_2

is essential for respiration of fresh fruits and vegetables, CO_2 is a bacterial and fungal growth inhibitor, nitrogen is inert, tasteless and less prone to pass either into the product or out through the packaging material. It forms an important filler gas.

For a fruit to be compatible with the use of CA it should preferably be characterized by : (i) a long postharvest life (ii) resistance to chilling injury (iii) a large range of non-injurious atmospheres (iv) resistance to fungal and bacterial attack (v) adaptation to a humid atmosphere (vi) a climacteric fruit that can be ripened during or after storage (vii) absence of negative CA residual effect.

4.4 Benefits and Harmful Effects of MA and CA

4.4.1 Beneficial Effects

When used properly, MA and CA can supplement proper temperature management and can result in one or more of following benefits:

i) Retardation of senescence (ripening)

ii) Slowing down of respiration and ethylene production rates

iii) Retards softening and compositional changes

iv) Reduction of fruits sensitivity to ethylene action.

v) Alleviation of certain physiological disorders such as chilling injury of various commodities, russet spotting in lettuce, and some storage disorders of apple.

vi) Have direct or indirect effect on postharvest pathogens and consequently decay incidence and severity.

vii) Modified atmosphere can be a useful tool for insect control in some commodities.

viii) Can replace postharvest chemicals used for control of some physiological disorders. Fungicides and fumigants may be reduced or eliminated.

ix) In specific instances, facilitate picking and marketing more mature (better flavoured) fruits by slowing down their postharvest deterioration.

4.4.2 Harmful Effects

i) Initiation and/or aggravation of certain physiological disorders such as blackheart in potato, brown stain of lettuce, and brown heart in apple and pear.

ii) Irregular ripening of fruits such as banana, pear and tomato at O_2 levels below 2 per cent and CO_2 levels above 5 per cent.

iii) Development of off-flavors and off-odors.

iv) Increased susceptibility to decay when the commodity is physiologically injured by too-low O_2 or too-high CO_2 concentrations.

v) Stimulation of sprouting and retardation of periderm development in some root and tuber vegetables.

vi) MA and CA can be injurious and even deadly to humans entering a room container without proper safety equipment.

vii) MA and CA can cause structural damage to rooms and containers that lack proper pressure relief systems.

4.5 Modern CA Storage System

Modern CA storage system is much safer than the traditional ones due to the use of the following systems for N_2 generations.

4.5.1 Use of Pressure Swing Adsorption (PSA) for Nitrogen Generation

PSA nitrogen gas generator has two sets of columns filled with carbon molecular sieves (CMS). Under Pressure these columns retain all active molecules compounds present in air like (Oxygen, Carbon Dioxide and Water) with the exception of the inert gases like Nitrogen and Argon. The N_2 gas can be produced onsite for direct use or can be stored in nitrogen storage tanks for intermittent use (Nitrogen and Argon). This process is referred as PSA (Pressure Swing Adsorption).

4.5.2 Use of Membranes for Nitrogen Adsorption

In this system compressed air is passed through the specialized membranes and series of filters. The membrane separates the N_2 form the air through exclusion of O_2. The compressed N_2 comes out form the other end which could be stored in tanks for future use.

4.5.3 Safety Considerations

The atmosphere in CA rooms will not support human life. People may die of asphyxia while working inside CA rooms without breathing apparatus. A danger sign should be posted on the door. The access hatch should be large enough to accommodate a person with breathing equipment strapped to his or her back. At least two persons with breathing equipment should work together at all times with one inside and one outside the room watching the first person.

5.0 Conclusion

In appropriate management of horticultural crop produce, first and foremost important priority is the estimation of losses. Mechanisms need to be developed for estimation of near accurate postharvest losses. Postharvest losses could be managed through interventions like choice of varieties, sorting, grading, use of varieties resistant to biotic and abiotic stress, good horticultural practices, market intelligence and storage.

Postharvest management practices that slow down the postharvest deterioration during handling and storage need to be practiced. This would be possible through improved capacity building of stakeholders including extension agencies, upscaling of technologies, developing linkages among people, private and public. Several policy issues for managing postharvest losses also need to be addressed. Amendment of contract farming rules, APMC acts, development of postharvest infrastructure, energy efficient storage system, crop insurance etc. are coherently linked with each other for achieving the desired goals for salvaging huge postharvest losses of horticultural produce in India

Among the fruits top priority needs to be given to postharvest management of apple, banana, citrus, guava, mango, papaya, pineapple, pomegranate, strawberry and sapota. The priority crops in vegetables are brinjal, cabbage, cauliflower, okra, onion, tomato, potato, sweet paper, chilli, garlic. Among them rose, carnation, gladiolus, tuberose, lilies, orchids, chrysanthemum should get priority attention. Ginger and Turmeric are the most important spice crops in India requiring priority.

Development of non-destructive methods to determine maturity of fruits and vegetables is need of the hour. Similarly, postharvest research should also focus on understanding fundamentals of fruit ripening including regulation of ethylene biosynthesis in tropical and subtropical fruits of Indian origin for developing commercially practicable technologies. Research on development of sound and economical handling technologies for specific fruit, vegetable, cut flowers, spices and other horticulture crops are some of the research priorities in India. Use of sensors for detecting infected/injured or damaged food including use of multiple chips, nano-biosensors and efficient grading and sorting of healthy produce are also essential. In case of root and tuber crops appropriate curing practices need to be standardized for improving the postharvest quality and reducing the losses during handling. Development of smart/intelligent packaging, punnet, CFBs, plastic trays, stretch films, shrink wraps recyclable plastic crates etc. for different types of produce are also emerging area of postharvest research.

Integrated diseases and pest management practices involving minimum chemical use imparting natural disease resistance against plant pathogens and use of bio-agents are some of the novel areas of research. Research on integrated insect pest and disease management in both pre and postharvest phases *i.e.* use of hormone traps for fruit flies, hot water dip, VHT, insecticidal controlled atmosphere, gamma irradiation etc. need to be attempted in order to meet quarantine requirements for overseas markets.

In the area of storage of horticulture produce, development of genotype-specific information on use of cold storage and CA/MA storage (conc. of O_2, CO_2 and temperature) are lacking. Similarly, multi-commodity storage needs to be developed for diversified produce. Low temperature storage through intermittent warming, CA, growth regulators improving the efficiency of surface coating and packaging are some of the priority areas. Development of edible and eco-friendly coating (nano-polymers, chitosan) for extending postharvest life also deserves attention. Combined use of computer and engineering tools to estimate precise cost benefit relationship among structural geometric, thermal and ergonomic aspects of commodity packaging and storage is another area.

Genotypes having resistance to physiological disorders and/or decays causing pathogens/pests need to be evolved to minimize the use of chemicals for better shelf life. Studies on genomics, transcriptomics, proteomics and metabolomics in relation to fruit maturity, physiology of ripening, senescence and susceptibility to diseases and pests need to be carried out. Development of nano-biodegradable membrane for smart packaging, nano formulations/molecules/particles for reducing the ethylene and colour development and nano coating/waxes for better retention of flavor/colour minimizing microbial spoilage are some of the future researchable issues.

Development of holistic quality management system involving safe production practices, harvesting, postharvest handling, storage, transportation, cold chain, distribution and assurance on safe food production is essential. Interdisciplinary approach is the need of the hour to tackle any postharvest problems involving horticulturists, postharvest biologist, plant breeders, molecular biologist, food technologists, microbiologists, bio-chemists, plant physiologists, and agri./food engineers, computer scientists, statistician, economists, management expert and psychologists.

6.0 Important References

Abeles, F.B., Morgan, P.W. and Saltvett, M.E. 1992. **In:** *Ethylene in plant biology*, 2nd edition, Academic Press, London, UK, pp. 182-186.

Chadha, K.L. 1995. Status of postharvest technology of fruits. In: *Horticultural Renaissance in India.* A compendium of selected lectures of Dr. K.L. Chadha (Eds.) S.M Paul Khurana and S.K. Singh, IARI, New Delhi, 2013, pp. 905-12.

Chadha, K.L., 2000. Pre and Postharvest technology of fruits and vegetables. In: *Horticultural Renaissance in India.* A compendium of selected lectures of Dr. K.L. Chadha (Eds.) S.M Paul Khurana and S.K. Singh, IARI, New Delhi, 2013, pp. 935-952.

Chesworth, J.M., Stuchbury, T. and Scaife, J.R. 1998. **In:** *An Introduction to Agricultural Biochemistry*, Chapman and Hall, London, UK, pp. 388-392.

HLPE, 2014. Food losses and waste in the context of sustainable food systems. A report by the *High Level Panel of Experts on Food Security and Nutrition* of the Committee on World Food Security, Rome.

Kader, A.A. 1986. Biochemical and physiological basis for effects of controlled and modified atmosphere on fruit and vegetables. *Food Technol.*, **40(5)**: 99-100, 102-104.

Kader, A.A., Zagory, D and Kerbal, E.L. 1989. Modified atmosphere packaging of fruits and vegetables. *Critic. Rev. Food Sci. Nutr.*, **28**: 1-30.

Lakshminarayana, S., Shetty, M.S. and Krishnaprasad, C.A. 1975. Accelerated ripening of Alphonso mangoes by the application of ethrel. *Trop. Sci.*, **17:** 95-101.

Lalel, H.J.D., Singh, Z. and Tan, S.C. 2003. Maturity stage at harvest affects fruit ripening, quality and biosynthesis of aroma volatile compounds in *'Kensington Pride' mango. J. Hortic. Sci. Biotechnol.*, **78,** 225-233.

Lesham, Y.Y., Halevy, A.H and Frenkel, C., 1986. Fruit ripening, **In:** *Process and Control of Plant Senescence*. Elsevier, Amsterdam, the Netherlands, pp. 162-201.

Ludford P.M. and Isenberg F.M.R. 1987. Brassica crops. **In:** Weichman J (ed) *Postharvest Physiology of Vegetables*, Marcel Dekker Inc., New York, pp 497-522.

Mahajan B. V. C., Kaur Tajender, Gill, M.I.S., Dhaliwal H. S., Ghuman, B. S. and Chahil B. S. 2010. Studies on optimization of ripening techniques for banana. *J Food Sci Technol.*, **47(3):** 315–319.

Mahajan B.V.C., Singh G. and Dhatt, A.S. 2008. Studies on ripening behaviour and quality of winter guava with ethylene gas ethephon treatments. *J Food Sci Technol.*, **45:** 81–84.

Medlicott, A.P., Sigrist M.M., Reynolds S.B. and Thompson A.K. 1987. Effect of ethylene and acetylene on mango fruit ripening. *Ann Appl Biol.*, **111**: 439–444

Nguyen, H.R., McConchie, Hofman, P., Smith, L., Stubbings, B. and Adkins, M. 2002.Effect of ethylene and ripening temperatures on the skin colour and flesh characteristics of ripe 'Kensington Pride' mango fruit. *Acta Hortic.*,**575,** 635-642.

O'Connor, Skarshewski P. and Thrower S.J. 1992. *ASEAN Food J.*, **7:** 127-136.

Pal, R.K. 1997. Pre-storage exposure to Controlled Atmosphere by using simple gas blending system for better storage of mango. **In:** *Proc. of symposium on Processing, finance and marketing in food industry* (H.O.Gupta, Ed.), AFST (I), Delhi Chapter, October, 1997, pp.96-104.

Pal, R.K. 1998. Influence of ethrel and calcium carbide on respiration rate, ethylene, evolution, electrolyte leakage and firmness of 'Dashehari' mango (*Mangifera indica*) in India. *J. Agric. Sci.* **68:** 201-203.

Pal, R.K. and Buescher, R.W. 1993. Respiration and ethylene evolution of certain fruits and vegetables in response to carbon dioxide in controlled atmosphere (CA) storage. *Journal of Food Science and Technology*, **30(1): 29-32.**

PFA, 2003 *Prevention of Food Adulteration Act* 1954 with Prevention of Food Adulteration Rules 1955. 19th edn, International Law Book Co, Delhi, p. 140.

Reyes M.U. and Paull, R.E. 1995. Effect of storage temperature and ethylene treatment on guava fruit ripening. *Postharvest Biol Technol.*, **6**: 357–365.

Saltveit, Mikal E. 1999. Effect of ethylene on quality of fresh fruits and vegetables. *Postharvest Biology and Technology*, **15**: 279–292.

Salunkhe,D.K. and Desai B.B. 1984. *Postharvest Biotechnology of Vegetables*. Vol. I, CRC Press Inc.

Thompson, A.K. and Seymour, G.B. 1982. Comparative effects of acetylene and ethylene gas on initiation of banana ripening. *Ann. Appl. Biol*. **101**: 407- 410.

Valverde, F., Hermadez, R.L. and Sain, M.V. 1986. Effect of ethephon and calcium chloride on the ripening and quality of mango fruit, cv. Kiett in Alajuela, Costa Rice. *Agronomica costarricense* (Costa Rica), **10(1-2):** 153-163.

Yang, S.F. 1980. Regulation of ethylene biosynthesis, *Hort Sci.*, **15**: 238–243.

Chapter 2

Fruit Bagging

Ram Asrey

Principal Scientist, Division of Food Science & Postharvest Technology,
ICAR-Indian Agricultural Research Institute,
New Delhi – 110 012, India
E-mail: ramu_211@yahoo.com

1.0 Introduction

Preharvest fruit bagging is one of the best available alternatives which address the concerns of the farmers, consumers and industry people about residue free quality produce. It is an effective tool to improve the fruit quality and commercial value of the fresh produce. The bag provides a physical protection to the fruit by preventing insect pests and damage due to external injury. For over 200 years, the Japanese tied little bags over developing fruits. They first used sewed white silk bags until the 1960's when plastic bags became readily available. Later on bagging became prevalent throughout Europe in pome fruit orchards during Nineteen Seventies (1970's) and Eighties (80's). Wrapping of fruits before harvest with newspaper is used in several Asian countries for fruit fly control.

Fruit bagging is commercially employed in apple, peach and pear world over. It is also used on a limited scale on late mango cultivars in south-east Queensland (Australia) and there is growing interest for its use in case of litchi. This approach is used to produce high quality, unblemished litchi fruits in South Africa for litchi. In India, fruit bagging is not commercially employed in any crop so far but has been gaining momentum in pomegranate, apple, litchi and mango. Bagging affects pigment synthesis, days to maturity, cell histology, physiology, shelf-life, storage behaviour, biochemical composition, pesticide residue content and other phytosanitary quality aspects of fruit. The impact of fruit bagging depends on several factors and cofactors like genotype, bagging material, colour, time of bagging and covering duration. While bagging has several benefits, its dependency on manual operation limits the practical feasibility of this technology. Therefore, there is a need to design and develop bagging gadgets/machinery to mechanize this operation.

Bagging affects produce quality through modification in solar light quality–UV rays and other visible rays; Change in microclimate (temperature and relative humidity) around growing fruits; Absorption and translocation of nutrients; Alteration in sink-source relationship; Alleviation of insect pest incidence. It also helps in reducing pesticide residues and advancing maturity in certain crops.

2.0 Bagging Materials and Application Time

Fruits can be easily protected by bagging them in newspaper, cellulose or butter paper bags. These bags are available in a range of colours *e.g.*, white, red, blue, orange etc. Plastic bags may also be used but these are not ideal due to, moisture and heat accumulation inside the bag that favours fungal growth. Alternatively, bags made of leaves like of like pandanus, betel nut tree may also be used. Different fruit crop behave differently on the use of particular bag colour. For example, bagging of peach gives desirable results with the orange colour bags while apple does better with red coloured bagging. The growers can select the bag colour on the basis of existing available information.

To bagging operation can be performed during sunny bright days when there is no moisture deposition on the fruit surface. Bagging may be done manually by standing on ground or with the help of a tripod ladder. Healthy looking fruit (individually or in bunches) are placed in the bag and its top end is tied up with string or a wire. The bottom of the bag is pushed upwards to make it 'V'-shaped. This prevents damage by rain and keeps the bag inflated keeping the fruit away from touching the sides of the bag.

Timing of fruit bagging is crucial for obtaining better results. Researchers have worked out the ideal bagging time for some commercially important fruit crops. The ideal bagging stage for selected fruit crops and its impact on postharvest quality attributes are given in Table 1.

In pome fruits, bagging is generally performed at an early stage when the fruits are in active growing stage. Tropical and subtropical fruits respond better when they are bagged 60 to 40 days before harvesting.

The following precautions need to be taken while fruit bagging

i) Insertion of leaves, twigs, flower buds and flowers inside the bags should be avoided.

ii) Only disease free and healthy looking fruits should be bagged.

iii) Both the ends should be properly closed to avoid sheltering of insect inside bags. However, for preventing sun scalding one end can be kept open.

iv) Bagging shold be done one day after fungicide spray during bright sunny weather.

Table 1: Impact of Fruit Bagging on Postharvest Quality Attributes

Crop	Fruit Bagging Stage	Impact on Quality	References
Apple	Early stage (4 to 6 weeks after petal fall)	☆ No impact on maturity ☆ Higher phenolics and anthocyanin content ☆ Low level of procyanidin and quercetin glycosides ☆ Higher anthocyanin retention during storage ☆ Lower Ca content leading to bitter pit incidence	Ju, 1998 Arakawa, 1998
Banana	After bunch formation	☆ Lower incidence of fruit speckling, anthracnose, cigar-end rot and scaring beetle ☆ Higher shelf-life, dry matter content and early maturity	Stover and Simmonds, 1991 Pasberg-Gauhl, 2000
Grape	70 days before harvest	☆ Uniform fruit colour ☆ Lower TSS, higher acidity and late maturity Higher sucrose and lower glucose and fructose level ☆ High hue angle and low chroma value	Signes *et al.*, 2007
Litchi	30-40 days before harvest	☆ Higher fruit retention and colour intensity ☆ No effect on TSS/acid ratio	Tyas *et al.*, 1998
Mango	100 days before harvest	☆ Lower incidence of anthracnose and stem-end rot ☆ Reduced fruit Ca content and higher N in fruit ☆ Higher dry matter content and early maturity	Hofman *et al.*, 1997
Mangosteen	14 days after full bloom	☆ Bagged fruits more attractive but no statistical difference in peel anthocyanin content ☆ Higher PAL activity (known for anthocyanin synthesis) in bagged fruits ☆ Prevents fruit fly infestation	Ratanamarmo *et al.*, 2005
Peach	50 days after full bloom	☆ Higher fruit length and firmness ☆ Higher TSS ☆ Bright red coloured fruit	Bin *et al.*, 2006
Pear	30 days after full bloom	☆ Reduction in the level of blemishes and friction induced discolouration on fruit	Amarante *et al.*, 2002
Pomegranate	80 days before harvest	☆ Higher juice and anthocyanin content ☆ Low incidence of bacterial blight and fruit fly	Asrey *et al.*, 2011

3.0 Effect of Bagging on Fruit Quality

3.1 Apple

It has been reported that covering of fruits with one, two or three-layered paper bags from early development stage until harvest did not affect fruit maturity but significantly inhibited anthocyanin accumulation in fruit peel compared with the un-bagged control of apple cv. Delicious (Figure 1). High levels of procyanidin, quercetin glycosides, and simple phenols were detected in fruits covered by one and two-layered bags. When fruits treated with three-layered bags were exposed to light, they started accumulation of anthocyanins rapidly and there accumulation reached maximum after 3 days of light exposure. The anthocyanin content in fruits remained constant during 5 months of cold storage.

Figure 1: Bagging of Apple Fruits

Poor red colouration is often a serious problem in the production of Fuji apples. To overcome this situation, the practice of fruit bagging is common in Japan. It has been reported that the bagging of apple fruits 4 to 6 weeks after petal fall developed a brilliant red or pink colour after removing the bags 4-6 weeks before harvest. However, the ability of the fruits to develop red colour is likely to decrease more sharply with advance in fruit ripening. Therefore, timing of fruit bagging has a paramount importance on overall fruit quality.

3.2 Banana

Generally bagging in banana is performed after bunch formation by covering with a large translucent perforated polyethylene bags to prevent peel blemishes and to create a greenhouse effect around the fruit by raising the temperature about 0.5 °C (Figure 2). Bagging helps the fruit to fill out and mature more quickly. The greenhouse effect can be intensified by reducing size and number of the perforations. However, if the holes are too small and not numerous enough, the ventilation of the bunch is reduced leading to an increased humidity level within the bag. Banana bags with 5 per cent vent holes of about 4 mm in

Figure 2: Bagging of Banana Bunches

diameter give best result. If early bagging is practiced, it is necessary to clean the bags to avoid the accumulation of fallen bracts and flower remnants. Fruit speckling, anthracnose, cigar-end rot and scaring beetle can be minimized effectively through bagging. Higher shelf-life, finger dry matter content and early maturity has been reported by several workers in bagged banana bunches.

3.3 Grape

Bagging of grape bunches provide uniform colour to the whole bunch because the grapes ripen in the protected condition against the direct exposure of sunlight. Bunches when covered with cellulose bags had significantly decreased level of soluble solid content and (soluble solid to titratable acidity) ratio of 'Perla' grape. The concentration of sucrose was found higher in bagged fruit while it contained lower level of glucose and fructose than un-bagged fruits. The organic acids, citric, malic and tartaric were found highest in bagged fruits compared to un-bagged fruits. Grapes covered with bags had higher values of lightness, hue angle and lower values of chroma than un-bagged fruit.

3.4 Litchi

Litchi cv. Tai So when covered with paper bags increased the retention of fruit per panicle as compared to control. It also increases fruit size reduces fruit water loss. Bagging litchi reduced the percentage of rejected fruit compared to no bagging. Bagging of panicles at 42 days before harvest showed the lowest percentage of rejected fruits. Longer bagging time increased the percent rejects. However, bagging had no effect on TSS, acidity or their ratio in litchi.

3.5 Mango

Bagging of Mango (cv. Keitt) with white paper bags, 100-days before harvest influences fruit quality and mineral composition (Figure 3). Bagging reduces the incidence and severity of *Anthracnose* and stem-end rot through a reduction in contact between disease propagules and fruit. Bagging treatment increased the percentage of skin area with yellow colour at the eating soft stage. However, the percentage of red colour and its intensity decreased with increasing duration of bagging.

Figure 3: Bagging of Mango Fruits

Fruit dry matter and Nitrogen concentration was reported to increase, with faster ripening in bagged fruits as against unbagged ones. Fruit mass, flesh colour, total soluble solids, acidity and eating quality were generally not affected by bagging.

3.6 Mangosteen

Mangosteen (*Garcinia mangostana* L.) is known for attractive purplish-red colour which influences consumer acceptance and sales. Fruit fly infestation, otherwise a serious problem in mangosteen could be successfully prevented by bagging. The major pigment associated with the attractive colour is cyanidin-3-sophoroside. Mangosteen attains harvesting maturity when pericarp is light greenish yellow scattering with pinkish spots. Bagged mangosteen fruits at 14 days after full bloom when harvested at commercial maturity stage (when pericarp background turned pinkish) exhibited attractive visual appearance with no statistical difference in peel anthocyanin content. Bagged fruits exhibited higher PAL activity (enzyme known for anthocyanin synthesis) and low TSS over unbagged ones.

3.7 Peach

Preharvest bagging of peach (*Prunus persica*) cv. Hujingmilu at 50 days after full bloom with orange paper bags of single, double and triple-layers improved fruit quality. Fruits covered with triple-layered bags showed higher fruit length and firmness than un-bagged fruits at full-ripe stage. Both values of L (brightness) and hue angle (h °, red to green) were the lowest in un-bagged fruit, indicating darker red colouration of fruit compared with bagged fruit. TSS concentration in juice was found higher in single-layer bagged fruit than in other bagged and un-bagged fruit at the full-ripe stage. However, no consistent difference was found in titratable acidity. Single and triple-layer bagged fruits had higher ethylene production rates than double-layer bagged and un-bagged fruits. Single-layer bagged fruit showed the highest level of dakua-decalactone, a main characteristic aroma of peach and total lactones at the firm-ripe and full-ripe stages. Studies on 'Hakuho' peach revealed that bagging of fruit with orange parchment paper bags 15 days before harvest resulted in higher visual quality than un-bagged fruit. Fruits covered with orange bags had the greatest radiation interception resulting in the highest hue angle. The surface of fruits covered with orange bags had the highest reflectance in the range of 400 nm–650 nm, indicating that the fruit appeared bright red and had a high L value, resulting in good visual quality.

3.8 Pear

Reduction in bird damage (from 28.4 per cent for un-bagged to nil for bagged fruit) and skin blemish (49.5 per cent for un-bagged to 33.4 per cent for bagged fruit) is one of the major advantages of fruit bagging in pear. Bagging also helps in developing resistance of the fruit skin to damage during handling and storage. Preharvest bagging of pear fruits (*Pyrus communis* L. 'Doyenne du Comice') with micro-perforated polyethylene bags 30 days after full bloom has been reported to increase the percentage of acceptable quality fruits for export from 27.2 (in unbagged) to 63.2 per cent. . Bagging reduced the level of blemishes that are mainly caused by fruit friction against leaves and thin twigs during periods of strong winds. When trees were sprayed with captan and azinphos methyl (non-systemic fungicide) on the 3rd day before harvesting, the bagged fruits had 50 per cent less residues compared to unbagged fruits.

Red Chinese sand pears (*Pyrus pyrifolia* Nakai) were covered by light-impermeable paper bags with different levels of light permeable liners during their early development stage followed by re-exposing them totally for 2 or 3 weeks accumulated largest amount of anthocyanin. Bagging treatments did not affect total soluble sugars, but decreased organic acids content in the fruit. With increasing time after the bag removal, the concentration of chlorophyll, carotenoid, flavonoid and total phenols were reported to get enhanced. After the bag removal, the sucrose contents and PAL activities increased gradually and correlation analysis revealed that they were highly correlated with anthocyanin accumulation in two cultivars. This study suggests that anthocyanin biosynthesis in red Chinese sand pears is a highly light dependent process modified by genotypes.

3.9 Pomegranate

Figure 4: Bagging of Pomegranate Fruits

Among the fruits, pomegranate is the most ideal crop for fruit bagging to due to its higher individual fruit weight , response to suture pigmentation (due to colour richness) and proneness to disorders (cracking, aril browning, sunburn). The popularity of pomegranate is increasing among growers and consumers in India due to high yielding potential, better keeping quality and high nutraceutical value of fruits. During last decade, pomegranate crop has witnessed tremendous growth potential even in non-traditional areas like Punjab, Himachal Pradesh, Karnataka and south western part of Haryana. Unlike other crops, pomegranate growers are facing several problems like fruit cracking, improper colour development, internal aril breakdown and bacterial blight. Pomegranate (*Punica granatum* L.) is most suitable crop for practicing pre-harvest fruit bagging because of its bushy nature (bagging can be performed by standing on ground or by using tripod ladder). For getting desirable bagging results pomegranate fruits should be bagged with white cellulose/ butter paper at least 85-90 days before harvesting (Figure 4). Zipped polyethylene bags of 15x15cm size may also be used for bagging. Bagging with zipped polythene bags is cheap, rain proof and convenient. Through bagging fruit quality was improved, pathological and physiological problems were also found to be reduced. Fruit sucking moth is a serious problem in pomegranate and the damage may range upto 100 per cent crop loss. It has been reported that 100 per cent protection can be achieved by bagging of fruits immediately after fruit set using non-woven polypropylene bags. It also improves the fruit gloss and overall acceptability. Since bagging may increase the severity of bacterial blight, it may be used for *Haste* or *Ambe Bahar* crops.

4.0 Conclusion

Nowadays good agricultural practices are gaining the popularity world over to produce quality fruits with least use of pesticides. Of several such practices, pre-harvest fruit bagging has emerged as one of the best tool. It is a physical barrier creation technique, which improves visual fruit quality by promoting peel colouration and reducing blemishes along with enriching internal fruit quality. Pre-harvest bagging of fruits can also reduce disease, insect-pest incidence, sunburn of the peel, fruit cracking, bird damage and pesticide residue load on fruits. Due to its beneficial effects, bagging has become an essential pre-harvest operation of apple, banana, grape, litchi, mango peach, and pear cultivation in countries like Japan, Australia, China and USA. Selection of right kind of bag colour and fruit bagging stage is crucial to harness the desirable effects of pre-harvest fruit bagging. As of now, fruit bagging is practiced in pomegranate and apple fruit crops in a very limited scale in India. Fruit bagging is a laborer intensive practice and much suits to the dwarf statured fruit crops like pomegranate, banana, apple peach and pear which could be attended by the workers while standing on the ground or with tripod ladders.

5.0 Important References

Amarante, C., Banks, N.H. and Max, S. 2002. Preharvest bagging improves packout and fruit quality of pears (*Pyrus communis*). *New Zealand J. Crop Hortic. Sci.*, **30**: 93-98.

Arakawa, O. 1998. Colouring of Fuji apples by bagging. IDFTA, Compact Fruit Tree, **31(2)**: 34-35.

Asrey, Ram and Pal, R.K. 2013. Pomegranate fruit-bagging enhances desirable fruit Characteristics, ICAR News, **19(1)**: 15-16

Asrey, Ram, Pal, R.K., Kumar, J. and Sharma, R.R. 2011. Get more by bagging pomegranate on trees, *Indian Horticulture*, (ICAR), **56(2)**: 12-13.

Bin, L.I., Hui-Juan, J. and Xiao-Meng, Z. 2006. Effects of fruit pre-harvest bagging on fruit quality of peach (*Prunus persica* Batsch cv. Hujingmilu). *J. Plant Physiol. Mol. Biol.* **32(3)**: 280-286.

Du, C.T. and Francis, F.J. 1977. Anthocyanins of mangosteen (*Garcinia mangostana* L.), *J. Food Sci.* **42**: 60- 67.

Ganry, J. 1975. Influence du gainage des régimes du bananier avec une housse de polyéthylène sur la température des fruits dans les conditions de Neufchâteau (Guadeloupe). *Fruits*, **30**: 735-738.

Hofman, P.J., Smith, L.G., Joyce, D.C., Johnson, G.I. and Meiburg, G.F. 1997. Bagging of mango (Mangifera indica cv. 'Keitt') fruit influences fruit quality and mineral composition. *Postharvest Biol. Technol.* **12**: 83-91.

Huang, C.and Lee, X. 2009. Effects of fruit bagging on colouring and related physiology, and qualities of red Chinese sand pears during fruit maturation. *Sci. Hort.* **121(2)**: 149-158.

Jaroenkit, T., Maichoo, S., Manochai, P. and Ratanamano, S. 2010. Effect of fruit bagging on chemical compositions and skin pigments of fresh longan (*Dimocarpus longan* Lour.). *Acta Hort.* **863**: 397-402.

Jia, H., Araki, A. and Okamoto, G. 2005. Influence of fruit bagging on aroma volatiles and skin colouration of 'Hakuho' peach (*Prunus persica* Batsch). *Postharvest Biol. Technol.* **35**: 61-68.

Ju, Z. 1998. Fruit bagging, a useful method for studying anthocyanin synthesis and gene expression in apples. *Sci. Hort.* **77**: 155-164.

Kitagawa, H., Manabe, K. and Esguerra, E.B. 1992. Bagging of fruit on the tree to control disease. *Acta Hort.* **321**: 870-875.

Pasberg-Gauhl, C. 2000. 'Fruit speckling' on bananas in the Atlantic zone of Costa Rica. BASF. Limburgerhof. Germany. 28pp.

Ratanamarno, S. Uthaibutra, J. and Saengnil, K. 2005. Effects of bagging and storage temperature on anthocyanin content and phenylalanine ammonia-lyase (PAL) activity in mangosteen (*Garcinia mangostana* L.) fruit pericarp during maturation. *J. Sci. Technol.*, **27(4)**: 711-717.

Signes, A.J., Burló, F., Martinez-Sánchez, F. and Carbonell-Barrachina, Á.A. 2007. Effects of preharvest bagging on quality of black table grapes. *World J. Agril. Sci.* **3(1)**: 32-38.

Stover, R.H. and Simmonds, N.W. 1991. Bananas, Longman, Harlow, England. 468pp.

Tyas, J.A., Hofman, P.J., Underhill, S. J.R. and Bell, K.L. 1997. Fruit canopy position and panicle bagging affects yield and quality of 'Tai So' lychee. *Sci. Hort.* **72**: 203-213.

Chapter 3

Maturity Indices

R.K. Pal
Director, ICAR-National Research Centre on Pomegranate, NH-9, Solapur-Pune Highway, Kegaon, Solapur – 413 255, Maharashtra, India
E-mail: rkrishnapal@gmail.com

P. Suresh Kumar
Senior Scientist, ICAR-National Institute of Abiotic Stress Management, Malegaon, Baramati – 413 115, Maharashtra, India
E-mail: psureshars@gmail.com

1.0 Introduction

Consumption of fruits and vegetables is important for human health because these foods are primary sources of some essential nutrients and contain phyto-chemicals that may lower risk of various chronic diseases. The desirable characteristics of a fruit or vegetable are dictated largely by the market for which it is intended. Traditionally, the fresh market is the one to which the grower turns up to secure an immediate return on the sale of the produce. Various stakeholders in the produce handling system demand different attributes from the same produce. To producers a given commodity must have high yield, good appearance, easy to harvest and withstand long distance market. Whereas, appearance, firmness and shelf life are an important consideration from the point of view of wholesale and retail markets. Consumers are interested in freshness of the produce with good value for money. The quality of fruits and vegetables after harvest cannot be improved but it can be assured when harvesting is done at proper stage of maturity. Immature fruits when harvested will give poor quality and erratic ripening.

Maturity of fruits and vegetables is an indication of the development of the crop as a marketable product. Selection of right stage of maturity for harvest has considerable influence on storage life, quality and final acceptance of the produce by the consumer. If the produce is to be shipped to distant markets or stored for a better price, it should be picked at mature but unripe stage. However, unlike ripening stage, differentiation between premature and mature stage is difficult to detect. Since no prominent changes in firmness or colour are evident, often harvest indices become arbitrary and subjective. Understanding the meaning and measurement of maturity is therefore, key postharvest management while maturity is the stage of fruit development, which ensures attainment of maximum edible quality at the completion of ripening process. Therefore, maturity indices are important for deciding when a given commodity should be harvested to provide some marketing flexibility and to ensure the attainment of acceptable eating quality for the consumer. This article deals in detail with the importance of maturity in commodity handling system and discusses maturity indices in use for fruits and vegetables to satisfy different stakeholders of the chain.

2.0 Maturity

The principles dictating the stage of maturity of a fruit or vegetable for harvesting are crucial to its subsequent storage and marketable life and quality. Immature fruits are more subject to shriveling and mechanical damage and are of inferior quality when ripe. Overripe fruits are likely to become soft and mealy with insipid flavour soon after harvest. Any fruit picked either too early or too late in its season is more susceptible to physiological disorders and has a shorter storage-life than fruit picked at proper maturity. Mostly fruits are harvested when they have ripened on the tree. However, some fruits are picked mature but unripe so that they can withstand the postharvest handling when shipped to long distance. Most currently used maturity indices are based on a compromise between those indices that would ensure the best eating quality to the consumer and those that provide the needed flexibility in marketing. For export purpose, mandatory quality standards for most fresh fruits and vegetables include objective indices of maturity to ensure minimum acceptability of their flavour quality to consumers. Examples of these indices are soluble solids concentration (cherry, grape, grape fruit, kiwifruit, mandarin and pear), TSS: acid ratio (citrus fruits, pomegranate and grape) and flesh firmness (apple, pear). Colour is used as the index of maturity for apricot, nectarine, peach, persimmon, plum, raspberry and strawberry fruits.

Postharvest physiologists distinguish three stages in the life span of fruits and vegetables: maturation, ripening and senescence. Maturation is indicative of the fruit being ready for harvest (Figure 1). At this point, the edible part of the fruit or vegetable is fully developed in size, although it may not be ready for immediate consumption (FAO, 2001). Ripening follows or overlaps maturation, rendering the produce edible, as indicated by taste. Senescence is the last stage in the ontogeny of the plant organ, characterized by natural degradation of the fruit or vegetable, as in loss of texture, flavour, etc.

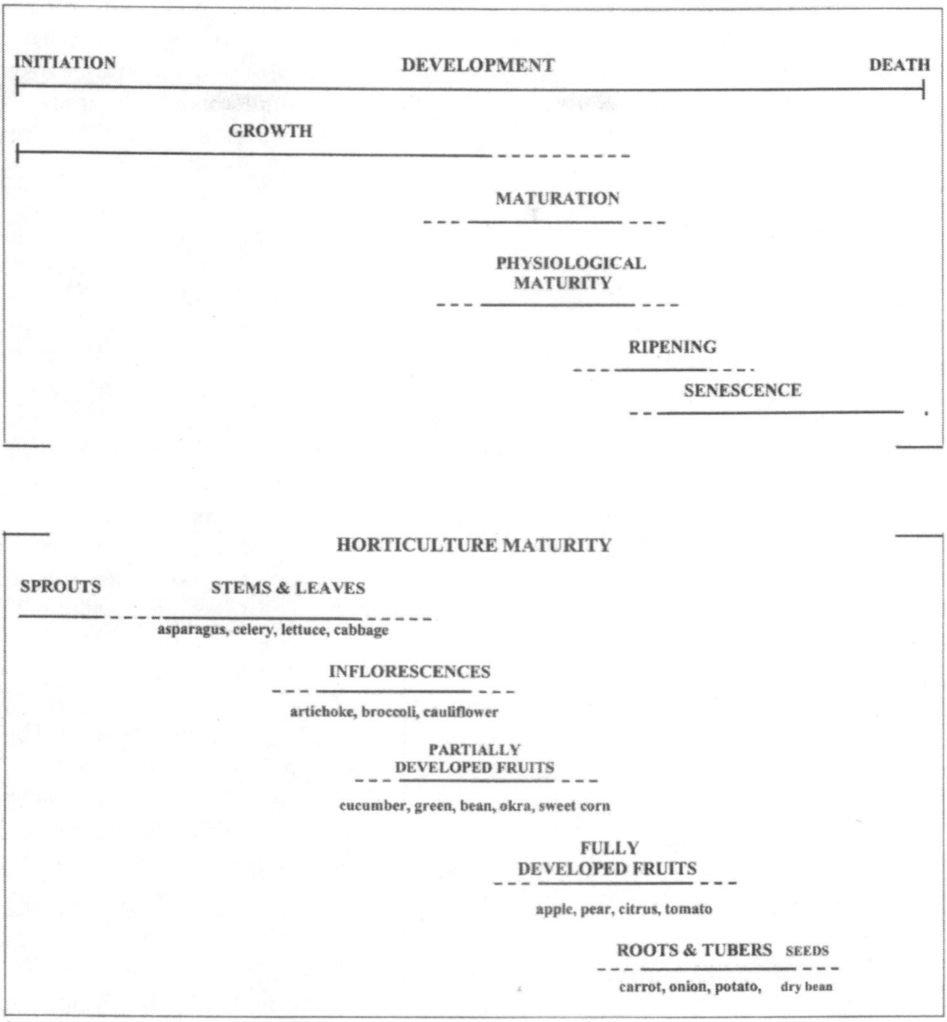

Figure 1: Horticultural Maturity in Relation to Developmental Stages of Plant (Watada *et al.*,1984)

2.1 Types of Maturity

Vegetables and fruits are harvested over a wide range of physiological stages, depending upon which part of the plant that is used as food. For example, small or immature vegetables possess better texture and quality than mature or over-mature for value addition etc. The maturity has been divided into broadly two major categories *i.e.* physiological maturity and horticultural maturity.

2.1.1 Physiological Maturity

It is defined as the stage at which the fruit attains complete maturation and becomes ready for further development of offspring or next generation. In other words it is the final stage of senescence. Although physiological maturity is most important index for maturity of cereals, pulses and oilseed crops, it is also important for certain vegetables in prediction of harvest maturity for their intended use. For example the physiological maturity of commodities *viz.* cucumber is attained when the seeds become hard and have acquired the capability of germination. Similarly, leguminous vegetables and okra are considered physiologically mature when seeds are fully developed and the pod will dehisce with little pressure.

2.1.2 Horticultural Maturity

Horticultural maturity emphasizes the stage of harvesting of fruits and vegetables as per the demand or acceptability by the consumer. Fruits and vegetables are to be harvested when plant or plant part possesses the prerequisites for the utilization by consumer for a particular purpose *i.e.* local or distant or export market (shipping maturity) or exhibition or processing (processing maturity), culinary maturity, dessert maturity. For instance, for local market and for processing, fully coloured tomato fruits are harvested while those which have just started developing colour *i.e.* turning or breaker stage are harvested for long distant marketing. Fruits are harvested at a proper stage of the maturity that exhibit excellent quality. Poor quality and uneven ripening are caused by early harvesting while late harvesting results in extremely poor shelf life. It is imperative that the physical damage to the fruit should be avoided at harvest apart from harvesting at appropriate stage of maturity for quality assurance during marketing.

2.2 Maturity Indices

Maturity indices are important for deciding when a given commodity should be harvested to provide some marketing flexibility and to ensure the attainment of acceptable eating quality to the consumer. Generally, a single maturity index is not considered to be reliable. In most of the crops, more than one or two indices should be made use of while determining the exact stage of harvest maturity. Fruits picked at the wrong stage of maturity may develop physiological disorders in storage and may exhibit poor quality. Fruit size is also sacrificed by harvesting too early.

After maturity of any organ it starts its decline stage *i.e.* called as "Ripening". Earlier the harvest, longer is the time of ripen. Greater the maturity, lesser are the number of days required for the fruits to ripen. Maturity indices are often called as "Maturity Standards" or "Signs of Maturity" for better understanding.

2.2.1 Developing Maturity Index

Some of the common strategies for developing maturity indices are:

i) Quantification of the changes in the commodity throughout its development

ii) Identification of physical features (size, colour, etc), whose changes correlate well with the stages of the produce development

iii) Conducting storage trials and organoleptic evaluation to determine the value of the maturity index that defines minimum acceptable maturity.

iv) An index value can be assigned for the minimal acceptable maturity after the relationship between changes in the maturity index of produce in quantity and quality over the storage life is determined, *v*) Examining the maturity index over several years and in multiple locations is prerequisite to standardize the right maturity so that farmers harvest better quality produce.

2.2.2 Maturity Standards

Many features of fruits and vegetables have been used to identify the right maturity indices for various commodities. The wide range of methods that have been devised to measure these features are listed below (Table 1 and 2).

Table 1: Major Features Used in Fruits and Vegetables to Determine Maturity

Index	*Example*
	a) General Characteristics
Days from anthesis	Melons, pineapple
Days from full bloom	Mango, citrus, apple and pear
Days from fruit set	Banana (90 days for Dwarf Cavendish group); mango (110-125 days for Alphonso and Pairi)
Development of abscission layer	Melon, apple
Drying of plant parts	Onion, garlic, banana, potato and ginger
Ease of separation	Muskmelon, grape and mango
Ease of snapping	Beans, okra and peas
Elapsed days (bloom to harvest)	Apple, pear
Heat units	Pea, date, apple, sweet corn, mango, grape, apple and pear
Leaf changes	Potato, bulb crops
Tapping	Watermelon and jackfruit
	b) Surface Characteristics
Cuticle formation	Grape, tomato
Netting	Musk melon
Peel colour	Citrus, papaya, pineapple, tomato, grape, mango, strawberry and pea
Pulp colour	Tomato, mango, apple
Skin colour	Rambutan (colour of spintern)
Specific gravity	Mango (1.0-1.02 for Alphonso; less than 1.0 for Dashehari), cherry, water melon, potato, pineapple (0.98-1.02), guava (1.00)
	c) Shape
Angularity	Banana
Compactness	Broccoli, cauliflower, cabbage, Brussels sprouts
Flattening of eyes with slight hollowness at centre	Pineapple

Contd...

Table 1–*Contd...*

Index	*Example*
Flattening of tubercles	Litchi
Fullness of cheeks	Mango
Size	Apple, asparagus, citrus, cucumber and pear
Solidity	Brussels sprout, cabbage and lettuce
	d) Textural properties
Firmness	Apple, pear, stone fruits
Internal colour	Tomato
Tenderness	Pea
	e) Compositional factors
Acidity	Citrus (Mandarin-0.4 per cent, sweet orange 0.3 per cent), mango and pineapple (0.5-0.6 per cent)
Aroma	Guava (β-caryophylene), jackfruit
Astringency	Persimmon, date
Ethylene	Apple, pear
Juice	Citrus (35-50 per cent)
Juiciness (milk exudation)	Sweet corn
Oil	Avocado, olive
Starch content	Apple, pear and banana
Sugar: acid ratio	Citrus, papaya, pomegranate, kiwi, melon
Sugar content	Apple, pear, grape, melon
TSS	Grape (14-16 °brix for Anab-e-shahi, 18-22 °Brix for Thompson Seedless, 12-14 °brix for Bangalore Blue), mandarin 12-14 °brix, sweet orange-12 °brix; papaya 11.5 °brix, pineapple–12-14 °brix.

Source: Kader, 2002; Dhatt *et al.,* 2007; Jadhav and Sharma, 2007.

Table 2: Maturity Indices of some Important Fruits and Vegetables

Crop	*Maturity Index*
	a) Fruits
Banana	Disappearance of angualirty in a cross section of the finger
Ber	Colour break stage (light yellow colour appear)
Cherry	TSS: 14-15 per cent, light red colour
Grapes (Table)	Minimum SSC 14-18 per cent ; SSC: TA of 20 or higher
Guava	When skin colour changes from dark green to light gren
Jack fruit	Tapping gives metallic sound, stopping of milk latex, outer skin turns blunt
Kinnow	TSS: acid ratio 12-14
Kiwi fruit	TSS 6.5 per cent ; firmness = 14lbs

Contd...

Table 2–*Contd...*

Crop	Maturity Index
Lemon	TSS: acid ratio; 30 per cent or more juice by volume
Litchi	TSS: acidity ratio 30-40, bright red in colour
Mango	Increase fullness of cheeks or bulge of shoulders, flesh colour change from yellow to yellowish orange
Papaya	Yellow skin
Peach	Ground colour changes from green to yellow
Plum	Change in skin colour
Pomegranate	The calyx gets curved inside and becomes hard and dry at maturity. Brix-acid ratio range from 25-40
Strawberry	2/3rd of berry surface showing pink or red colour
	b) Vegetables and Spices
Broccoli	Adequate diameter, compact, all florets should be closed
Cabbage	Compactness
Cantaloupe	3/4th to ful slip under slight pressure, abscission from vine
Carrot	Immature, roots reached adequate size
Cauliflower	Compact curd, at least 6" in diameter
Brinjal	Full size, firm and green but still tender, 40 DAS
Ginger	8-9 months after planting
Lettuce and Celery	Big size
Lima bean, Pigeon pea	Well filled pods, pods loose their greenness
Melon	Ground colour changes to white with greenish tint, slightly waxy peel
Mushroom	Caps well rounded, partial veil completely intact
Okra	Full size, optimum maturity can be judged by snapping the tip of fruit, pod 2-4", not fibrous, tips of pods pliable
Pea	Pods wel filled but not faded in colour
Pepper	Fruit size and colour
Potato, Onion and Garlic	Tops beginning to dry out and topple down
Watermelon	Dropping of tendril, change of fruit colour, full size, hollow sound when tapped. TSS: 10 per cent

Source: Kitinoja and Gorny, 1998.

2.2.3 Methods for Determination of Harvest Maturity

There are five types of indices to judge the maturity of the fruit.

a. Visual Means

Appearance, Skin colour, size, shape, persistence of style portion, drying of outer leaf, drying of whole plant body, change in smell or flavour, dropping down of ripe fruits, texture.

b. Physical Means

Firmness, ease of separation from pedicel due to formation of abscission layer, specific gravity, weight of the fruit, lenticel number, surface morphology, weight.

c. Chemical Analysis

TSS, acids, starch, sugar, oil content, sugar-acid ratio, bio-electrical conductance, starch iodine test, tannin content, juice content, etc.

d. Computation Methods

Number of days for harvesting fruits from fruit set till maturity, calendar date, mean heat units (°Days), etc.

e. Physiological Method

Respiration rate, internal ethylene evolution.

2.3 Typical Maturity Indices

Maturity indices must be set specifically for each region and variety. Some of these are described in the following pages.

2.3.1 Size and Shape

The visual appearance of fruit and vegetable is the most important quality factor, which decides its price in the market. The consumer (wholesaler or retailer) observes the quality of fresh fruits and vegetables on the basis of their visual or external appearance. The produce should therefore be allowed to attain proper shape and size. Size is generally of limited value as a maturity index in fruit, though it is widely used for many vegetables marketed early in their development. Size is often specified as a quality standard, with large size generally indicating commercial over-maturity and under-sized produce indicating an immature state. The assumption, however, is not always a reliable guide for all-purpose. Therefore, objective evaluation of size and shape of the produce should be conducted to meet the desired quality characteristics. For banana, the width of individual fingers can be used to determine harvest maturity. Fruit shape may be used in some instances to decide maturity. For example, the fullness of cheeks adjacent to pedicel may be used as a guide to maturity of mango and some stone fruits. Some cultivars of banana become less angular in cross section as development and maturation progress. Mangoes also change shape during maturation. As the mango matures on the tree the angle between the shoulders of the fruit and the point at which the stalk is attached may change.

2.3.2 Abscission

Abscission zone develops on the stalk (pedicel) that attaches the fruit to the plant in many fruits during the later stages of maturation and the commencement of ripening. The abscission zone permits the fruit to get detached from the mother plant. Formation of abscission layer is considered as the maturity index for melon, avocado and bael (*Aegle marmelos*) fruit.

2.3.3 Surface Colour

Colour of the skin is commonly applied to fruits, since it changes with the fruit maturity and ripening. Assessment of harvest maturity by surface colour depends on the judgment of the harvester. There is initially a gradual loss in intensity of colour from deep green to lighter green and with many commodities, a complete loss of green colour with the development of yellow, red or purple pigments. Ground colour as measured by colour charts, is useful index of maturity for apple, pear and stone fruits, but is not entirely reliable as it is influenced by factors other than maturity. For some fruits, as they mature on the tree, development of blush *i.e.* additional colour superimposed on the ground colour, can be a good indicator of maturity. Examples are red or red-streaked apple cultivars and red blush on some cultivars of peach. Further, the surface colour of fruits and vegetables can also be measured as on objective evaluation using colour difference meter.

2.3.4 Aroma

Volatile chemicals are synthesized by most fruits as they ripen. Such chemicals give its characteristic aroma and can be used to determine the maturity or ripeness of fruit. These specific aromas may be detectable by human olfactory sense as well as through recent development of e-nose.

2.3.5 Leaf Changes

Quite often marked changes in the leaf quality determines the stage at which the vegetables should be harvested. In most of the root, tuber and bulbous crops, the condition of the leaves can likewise indicate the condition of the crop below ground. For example, if potatoes are to be stored, then the optimum harvest time is soon after the leaves and stems have died. If harvested earlier, the skins will be more prone to harvesting and handling damage and become susceptible to storage diseases. Similarly, harvesting of onion is adjudged when 75 per cent of the population exhibit bending of leaf tips that indicate the maturation of bulbs underground.

2.3.6 Firmness

Change in texture is another important index for determining the maturity standard of fruits and vegetables. Excessive loss of moisture may also affect the texture of crops. These textural changes are detected by touch, and the harvester may simply be able to gently squeeze the fruit and judge whether the crop can be harvested. Today sophisticated devices are available to measure texture in fruits and vegetables *e.g.* texture analyzers and pressure testers. A force is applied to the surface of the fruit, allowing the probe of the penetrometer to penetrate the fruit flesh, which then gives a reading on firmness. Commonly used pressure testers to measure the firmness of fruits and vegetables are the Fruit Firmness tester (Figure 2). A more elaborate test, but not necessarily more effective, uses instruments like the Instron Universal Testing Machine (Figure 3), Food Texture Analyzer etc.

2.3.7 Specific Gravity

Specific gravity is the relative gravity, or weight of solids or liquids, compared to pure distilled water at 62 °F (16.7 °C), which is considered unity. In practice, the fruit

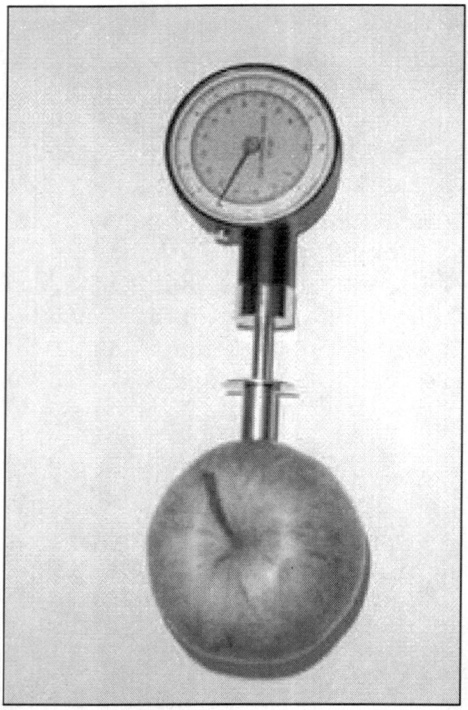

Figure 2: Fruit Firmness Tester

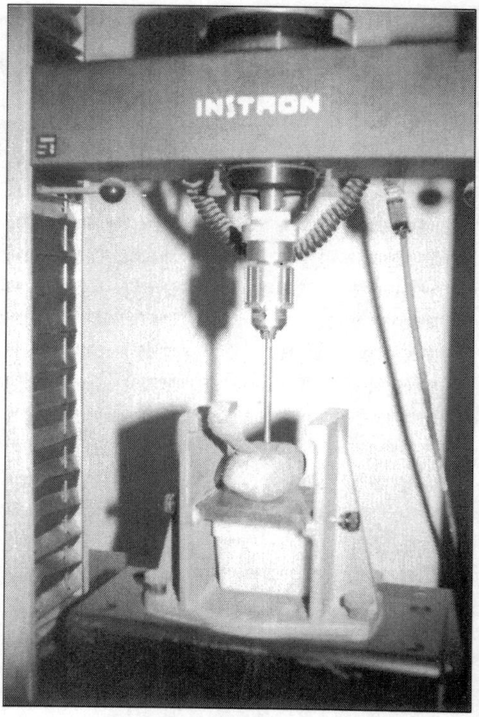

Figure 3: Instron Universal Testing Machine

or vegetable is weighed in air, then in pure water. The weight in air divided by the weight in water gives the specific gravity. This gives a reliable measure of fruit and vegetable maturity in many commodities *e.g.* Mango, Guava, Potato etc. As fruit matures, its specific gravity increases. Although this parameter is not practically used for determinining the harvest maturity, however, it is used to grade harvested produce according to different maturity grades.

Specific gravity of fruit is proved to be related to their internal characteristics such as dry matter, soluble solids or physical disorders. Commercially it has been used in mangosteen and potato. High specific gravity potatoes (above 1.089-1.080) are better suited for baking, frying, mashing and chipping while low specific gravity (1.079 – below 1.060) potatoes are suitable for boiling and canning. In case of watermelon, density of fruit is related to the degree of hollowness. The density of an immature watermelon is sometimes more than 0.98 g/cm^3. Watermelons with density lower than 0.934 g/cm^3 had cavities or unusually soft flesh of overripe porous tissue. Therefore, watermelons having density of approximately 0.95 g/cm^3 are considered to be optimum mature ones.

2.3.8 Juice Content

In case of citrus fruits, per cent juice content is one of the most important maturity indices. As per international standard no citrus fruit is considered mature if it contains less than 25 per cent juice by weight. Juice content of orange group fruits at optimum maturity ranges from 30-40 per cent whereas in case of limes and lemons the requirement is 25 per cent.

2.3.9 Oil Content

Oil content is used to determine the maturity of avocado fruits. Avocados at the time of harvest and at any time thereafter shall not contain less than 8 per cent oil in edible pulp by weight basis. The oil content is determined by weighing 5-10 g of avocado pulp and then extracting the oil with a solvent (*e.g.*, benzene or petroleum ether) in a distillation column.

2.3.10 Starch Content

Measurement of starch content is a reliable technique used to determine maturity in many fruits and vegetables. The cut surfaces stain to a blue-black colour in places where starch is present. Starch converts into sugar as harvest time approaches. Harvest begins when about 65-70 per cent of the cut surfaces turn blue-black. Starch content is used as maturity index for pea, potato, apple and other stone fruits.

2.3.11 Sugars

In climacteric fruits, carbohydrates accumulate during maturation in the form of starch. As the fruit ripens, starch is broken down into sugar. In non-climacteric fruits, sugar tends to accumulate during maturation. This factor is used in many parts of the world to specify maturity.

2.3.12 Acidity

In many fruits, the acidity changes during maturation and ripening, and in the case of citrus and other fruits, acidity reduces progressively as the fruits mature on tree. Normally, acidity is not taken alone for determination of fruit maturity but in relation to soluble solids it is termed the brix: acid ratio.

2.3.13 Total Soluble Solids

Total soluble Solid (TSS) is the content of total water soluble solids in the juice of the commodity and is one of the major handy measurement techniques for determination of maturity indices in many fruits and vegetables. This is also often termed as Soluble Solid Content (SSC). Fruits and vegetables primarily contain sugars, acids and water soluble polysaccharides as total water soluble solids. Sugars comprise the major ingredient in most of the fruits and vegetables in the TSS reading. Therefore, in general, high TSS value represents the high accumulation of sugars at the time of optimum maturity with the exception in cases where water soluble polysaccharides are present in considerable high amount *e.g.* Ber, Okra, Bael Fruit, etc. In such cases the high value of TSS does not conform to the values of high sugars. A simple instrument known as Hand or Pocket Refractometer is an important tool used for determination of TSS (Figure 4). Recommended TSS for determination of harvest maturity of different fruits is mentioned below (Table 3).

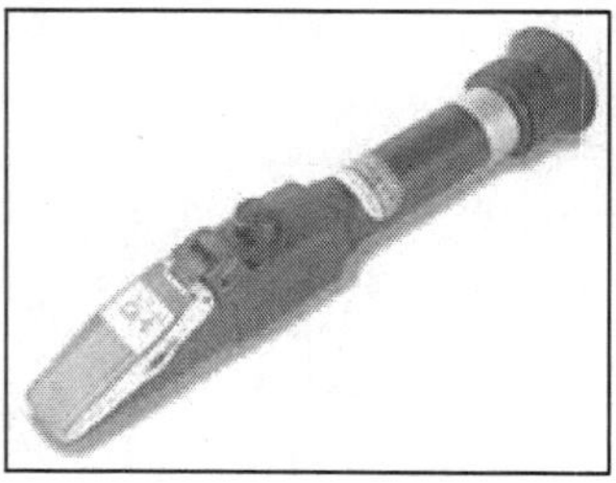

Figure 4: Pocket Refractometer

Table 3: Recommended TSS for Harvesting different Crops

Crop	*TSS (°Brix)*
Apple	10-12.5
Apricot, melon, nectarine, peach, water melon	10
Cherry, kiwi fruit	14-16
Grape	14-19
Grape fruit, mandarin, mango, orange, raspberry	8
Papaya	11.5
Pear	13
Persimmon	18
Pineapple, plum	12
Pomegranate	17
Strawberry	7

2.3.14 No. of Days/Heat Units

a) No. of Days

Time of flowering is largely dependent as temperature and the variation in number of days from flowering to harvest can be calculated for some commodities by use of degree-day concept. It has been observed that optimum maturity for harvesting of capsicum is judged at 40 days after flowering. Among the fruit crops banana fruit matures at 90- 110 days after fruit set whereas in case of mango it ranges from 110-130 days. In case of pomegranate the harvest maturity of cv. Bhagwa reaches between 170-180 days after flowering.

b) Heat Units

The number of degree days to maturity is determined over a period of several years by obtaining the algebraic sum from the differences, plus or minus, between the daily mean temperatures and a fixed base temperature (commonly the minimum temperature at which growth occurs). The average or characteristic number of degree-days is then used to forecast the probable date of maturity for the current year and as maturity approaches, it can be checked by other means.

This is an objective measure of maturity expressed as degree days or heat units in a particular environment. It has been found that number of heat unit or degree-days is required to mature a crop is almost remains constant. Hence accumulation of particular heat unit will require different number of days with respect to agro-climatic conditions in which the crop is grown. For this reason under warm conditions, maturity gets advanced and under cooler conditions, maturity gets delayed.

2.4 Detection of Maturity

With the advent of technologies, researchers and entrepreneurs use various non-destructive methods for identifying maturity of fruits and vegetables (Figure 5). Nuclear Magnetic Resonance (NMR), X-rays, ultra sonic waves, delayed light emission and

30 Days after Flowering **40 Days after Flowering** **50 Days after Flowering**

Figure 5: Different Stages of Maturity in Capsicum

(Source: Behera *et al.*, 2004)

light reflectance, Near Infra-red Spectroscopy (NIR), hyperspectral imaging technology, electronic nose, are some of the recent techniques which have vast potential to be used in future. The internal quality of fruit (maturity, firmness, soluble solid content, water content) could be detected with hyperspectral imaging system. Reflectance measurements could also serve as basis for prediction of ripening stages, estimation of harvest window and duration of storage for various fruits. Attempts to apply non-destructive methods based on optical spectroscopy for assessment of fruit quality and their physiological state have been undertaken for several decades. Portable fiber-optic radiometers providing reliable spectral data from very small fruit surface area for monitoring fruit maturity and quality have also gained considerable momentum. Researchers have demonstrated the use of the e-nose for monitoring ripening in a number of fruits including apple, banana, blueberry, grape, mandarin, peach and tomato.

Colour is the human visual perception of light reflected, transmitted, or emitted from an object in the visible portion of the electromagnetic spectrum from 380nm to 780nm.The main factor in the distribution of light energy reflected from the fruit is the presence and concentration of pigments including carotenoids, anthocyanins and other flavonoids, betalains, and chlorophylls in the skin. The colour aspect of visual appearance of the skin can be measured nondestructively using three types of sensors: colourimeters, spectrophotometers, and colour machine vision systems.

3.0 Recommendations for Harvest Maturity

3.1 Fruit Crops

3.1.1 Apple

Apart from attainment of desirable colour and size the TSS, starch content and fruit firmness are widely used as maturity standards for harvesting of apples. A simple iodine test can be used to judge the maturity of apples (Figure 6). A modified scoring pattern of iodine test for Indian apples have been developed for determination of harvest maturity. The starch-iodine test scores starting from 1 and ending at 4.5 over several weeks during the season. Harvesting of apple depends mainly on its

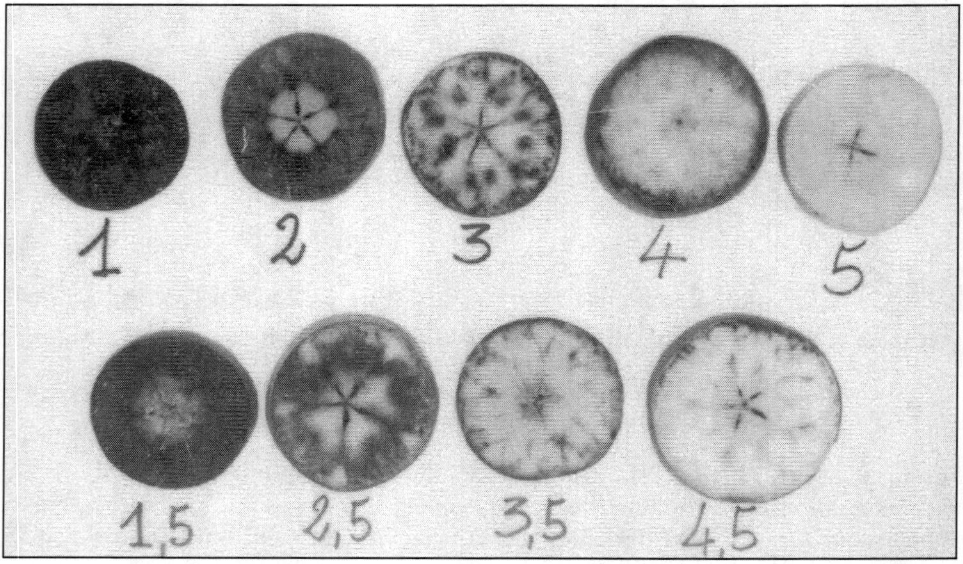

Figure 6: Starch-Iodine Score for Apple Maturity

market destination and type of storage is used. Apples destined for long-term controlled atmosphere (CA) storage have to be picked early between 2 and 2.5 with minimum TSS of 10 ° Brix and fruit pressure of approximately 18 lbs. Apples destined for immediate fresh market and processing plant can be picked later and therefore require a higher starch-iodine test score. The selection of the proper score largely depends on each individual operation. No single fixed score can be recommended to guide the whole industry.

The fruits mature within 130-150 days after the full bloom stage depending upon the variety. Time/temp (heat units) from anthesis, days from harvest to onset of ethylene production, ground colour, starch disappearance pattern, internal ethylene concentration and changes in firmness are used as maturity index for apple. The fruits at the time of harvest should be uniform, firm and crisp.The colour of the skin at maturity ranges from yellow to red depending on the variety.

3.1.2 Banana

Maturity of banana can be judged by degree of fullness of the fingers, *i.e.*, disappearance of angularity in a cross section (Figure 7). Harvest maturity of banana can be decided on the basis of market destination. Normally disappearance of 75 per cent or more angularity of

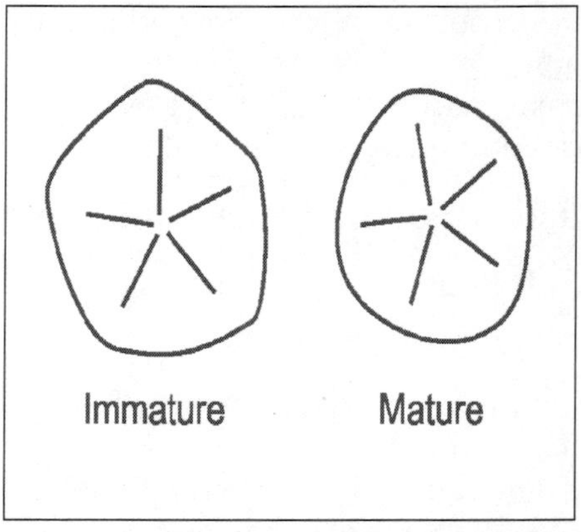

Figure 7: Disappearance of Angularity of Banana with Advancement of Maturity

fingers (Full three quarter stage) is considered as the optimum maturity index. When bananas are to be sent to distant urban markets they are best harvested in hard mature but unripe green state, which reduces the risk of deterioration during transport. Bananas are harvested mature-green and ripened upon arrival at destination markets since fruits ripened on the plant often split and have poor texture. The Dwarf Cavendish bananas are ready for harvest within 11- 14 months after planting, while tall cultivars take about 14-16 months to harvest. A bunch usually takes 90-120 days to mature after flowering, depending on climate and cultural practices. The maturity of banana is indicated by drying of top leaves, change in colour of fruits from dark green to light green and tendency of the floral end of the fruit to fall by slightest touch by hand. The dried parts of flowers at the top of fruit drop off easily. The mature fruit becomes plumpy and all the angles are filled in completely.

3.1.3 Blueberry

Colour and soluble solid-acid ratio are maturity indicators for blueberries. Fruits should be blue in colour without any green and soluble solids of 10 to 15 percent and pH 3.43-3.73. Pickers should handle fruits very carefully, by not holding more than a few berries in their hand at a time and use only low shallow containers so that fruits are not bruised due to piling.

3.1.4 Cashew

Cashew apple fruit matures in about 95 to 120 days from flowering. Depending upon variety apple colour changes from green to yellow or red whereas, nut becomes hard ash colour.

3.1.5 Cherry

Skin colour and soluble solids are the most common maturity indicators for sweet cherries. The soluble solids should be at least 16 °Brix and the characteristic skin colour for the variety can range from yellow to black red. Colour is the best indicator for sour cherries - these should be bright red.

3.1.6 Custard Apple

Custard apple fruits require 100 to 120 days for full maturity. The scales on fruits become prominent, plump and well spaced during optimum maturity. Fruit segments turn creamy white.

3.1.7 Date Palm

Dates ripen in four stages, known throughout the world by their Arabic names: *kimri* (unripe), *khalal* (full-size, crunchy),*rutab* (ripe, soft), *tamar* (ripe, sun-dried). In India, these stages are called gandora, doka, dang, and pind, respectively. Pind stage rarely occurs in India in India due to commencement of rainy season at that stage. Fruits from most of cultivars should be harvested at doka stage both for raw consumption as well as for its better keeping quality. The days taken from pollination to different stages of maturity and heat summation from spathe opening to maturity are also some important considerations for judging maturity in dates. Apart from it, significant reduction in tannin content coincides with the maturity of date palm.

3.1.8 Fig

In fig, the brown scurf on the fruit surface is replaced by yellowish corky brown colour at maturity. The absence of green tissues and latex also indicates maturity. Skin colour changes from light-brown with a tinge of green to light-brown to dark-brown. Flesh colour change from green to pinkish brown to reddish brown in some varieties.

3.1.9 Grape

Development of desirable Total Soluble Solids (TSS) Concentration of 14 to 17.5 per cent depending on cultivar and production area is the most reliable maturity index of grapes. In some situations, the TSS : acid ratio of 20 or higher is used to determine maturity for early maturing cultivars from early production areas. For red and black coloured cultivars, there is also a minimum colour requirement. The proper stage for harvesting the grapes is adjudged by the development of characteristic colour of a variety and also by feeling the grapes from the apical portion of the bunch. In grapes three major colours, *viz*. white, red and black are found. The criteria used to judge ripening differs according to the uses of grapes. For making raisins, harvesting at a late stage is preferred, to get more sugar for increasing the weight of the dried product. For all other purposes, ripening is judged on the basis of sugar: acid ratio for getting proper blend required for table purpose or wine making. The correct blend of sugar acid ratio is considered between 25-30. Grapes normally mature within 120 to 135 days after flowering. Easy separation of berries indicates harvesting stage. In addition, mature fruits develop characteristics flavours and aroma. In Anab-e-shahi, 15 to 16 per cent TSS is right time of maturity while for Thompson Seedless 18 to 20 per cent T.S.S is the optimum stage for harvest.

3.1.10 Guava

Guava fruits are picked at mature-green stage (colour change from dark- to light-green) where consumers eat them at that stage. Where consumers prefer ripe guava, the fruits are picked at the firm-yellow to half-ripe (softer) stage for long-distance transport or at the fully-ripe (yellow and soft) stage for local markets. Guava fruits ripen five months after flowering and are harvested when the fruits turn greenish yellow. Guava fruits are considered as fully mature when fruits become soft, develops slight depression near the stalk end, ridges on the fruit disappear and it becomes rounded. Mature fruits of guava become lighter in weight with specific gravity less than one. At this stage the force of detachment of the pedicel also becomes low.

3.1.11 Jackfruit

Jackfruit matures in about 180-270 days after flowering. The skin colour of the fruit changes from green to pale green. In mature fruits, the fruit spines become well developed and wider spaced with typical bluntness. A dull, hollow sound is produced when the ripe fruit is tapped by the finger. Though not a common maturity index, in some varieties the last leaf of the peduncle turns yellow with characteristic development of aroma.

3.1.12 Jamun

Fruits require 120-140 days to mature after flowering. In general the colour of the fruit changes from green to deep purple or black after attaining maximum size depending upon the variety.

3.1.13 Karonda

Karonda fruits generally require 87 to 90 days to mature after flowering. The colour changes from light green to dark green when the fruits are ready for harvest with optimum maturity. Red blush develops on the fruits of some types. Specific gravity just above 1 (1.012) is also considered as the indicator for maturity along with T.S.S. of 13.5 per cent, total sugar of 2.35 per cent and acidity of 2.62 per cent.

3.1.14 Kokum

Kokum fruits generally require 90-100 days to mature after flowering. The colour of the fruits changes from green to light green and red and purplish red when ripe. Average weight of fruit varies between 30-35 g depending on the type with a specific gravity of 1.04 at maturity. Chemical parameters like T.S.S. 13.68 per cent, acidity 3.61 per cent and total sugar 6.35 per cent are considered optimum for judging the maturity.

3.1.15 Litchi

Litchi fruits should be harvested when fully-ripe as they do not continue to ripen after harvest. The fruits for local market should be harvested at the full ripe stage as indicated by the attractive skin colour while for long distant market the fruits are harvested slightly early, when they have just started turning reddish or pinkish. Red of the fruit (due to anthocyanins in the skin) is a good indicator of maturity along with proper fruit size (minimum of 25 mm in diameter). Similarly the maturity of the fruit for harvesting is judged from the flattening of tubercles on it.

3.1.16 Mandarin and Sweet Orange

Citrus fruits are non climacteric, therefore, do not ripen further after harvest. To reach their full flavour and sweetness they must be left on the tree to ripen. In the tropics, fruits often remain green though they are fully ripe internally and do not develop an orange-yellow colour on the tree. The development of the orange-yellow skin colour (degreening) can be artificially induced after harvest. These facts make it very difficult to assess harvest maturity just from the appearance of the fruit on the tree. Some methods to identify the maturity are;

a) Skin Colour

This will be a good guide to maturity where colour develops normally; if normal skin colour does not develop, maturity may be indicated by a change in the shade of green shown by the skin; lemons change from dark green to a silvery-green appearance at maturity.

b) Size

Experienced growers may evaluate maturity by considering size as well as with other characteristics, such as slight changes in skin colour.

c) Internal Condition of Fruit

A few typical fruit thought to be mature are cut in to two, they can be considered ripe if:

i) Juice has developed full flavour and is sweet

ii) Fruit pulp has developed the normal colour

iii) Juice drips from the half-fruit when the cut surface is held vertically

iv) Desired TSS: Acid ratio

In general, citrus fruits mature in about 210 to 240 days. Colour changes from green to pale green or pale yellow or orange. In addition, outer rind looks shiny and one can see oil glands on the fruit. Mandarins have 0.4 per cent acidity and 12 to 14 per cent TSS whereas oranges are mature when acidity is 0.3 per cent and T.S.S is 12 per cent. Fruits in all citrus species become soft and glossy when they are fully mature.

3.1.17 Mango

Mangoes will not ripen normally if they are harvested before reaching the fully mature-green stage on the tree: they will lack sweetness and be poor in flavour. Mature-green fruit left on the tree will ripen and eventually fall. There are no simple reliable tests to indicate when mangoes are ready for harvest. A number of characteristics have been suggested for evaluating maturity, but they are not reliable for all cultivars or for all growing conditions. They must be interpreted in the light of local experience as below:

a) Fullness in Size

When fruits attain its full size, which mainly depends upon the variety, by virtue of experience one easily understands the final size and shape and hence its time of harvesting. It can be considered that mangoes are matured when the "cheeks" of the fruit are full. The shoulders of immature mangoes slope away from the fruit stalk; however, on more mature mangoes the shoulders become level with the point of attachment, and with even more maturity the shoulders may be raised above this point (Figure8).

b) Change in Peel Colour

Usually the colour of fruit peel changes from green to light yellow/apricot yellow. In some varieties, attractive red colour appears on shoulders (Alphonso, Kesar, Pusa Arunima and Arka Puneet). Sometimes, a variety may develop different colours in different localities. For instance, Dashehari in Malihabad region develops attractive yellow colour whereas, it remains green in Saharanpur region. Thus, harvesting has to be done as per size or colour it attains in a particular locality.

c) Natural Dropping of Ripe Fruits

Natural falling of ripened fruits from the tree is called as *tapka*. When some fruits start falling down it gives clear cut indication that mangoes are mature and must be harvested.

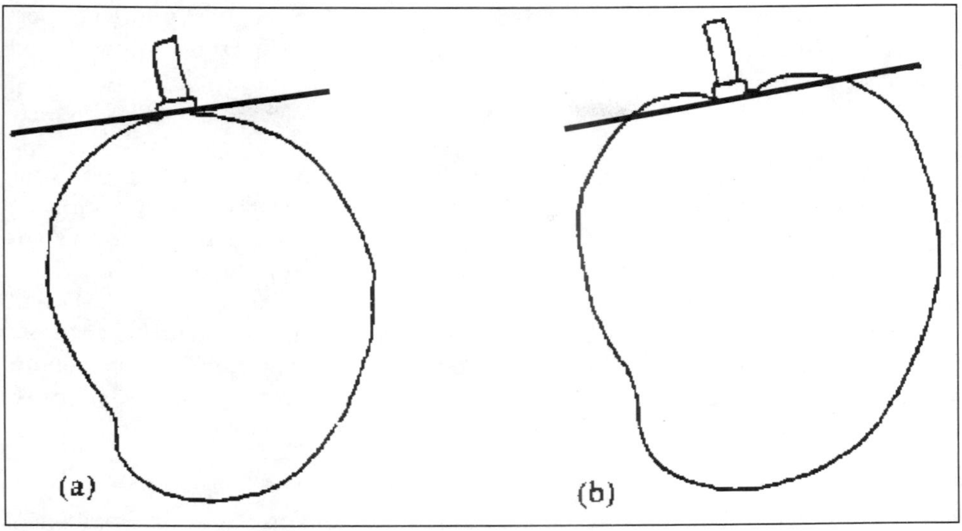

Figure 8: a) Immature and b) Mature Mango Fruits

d) Specific Gravity

This is scientific maturity index for harvesting mangoes. If the specific gravity of most of the fruits reaches between 1.01 and 1.02, it indicates that fruits are ready for harvesting. For this, some fruits are randomly harvested from the tree and put in a bucket full of water. Floaters are considered as immature and sinkers as mature (specific gravity > 1.02).

e) Number of Days to Maturity

It depends on the variety and climatic conditions. Usually, mangoes mature in about 90-140 days. Early varieties (Bombay Green) mature in about 90-100 days and late varieties mature in about 130-140 days (Amrapali, Pusa Arunima). In mature fruit, the sap flowing from the cut stem at harvest is thick and does not flow freely. Alphanso and Pairi take 110 to 125 days after fruit set for surface colour to change from dark green to olive green and flesh colour from white to pale yellow.

3.1.18 Mangosteen

The mangosteen fruit has an edible aril that is white, juicy, sweet and slightly acidic with a pleasant flavour, located inside a dark purple pericarp that is rich in bioactive secondary metabolites including anthocyanins, oligomeric pro-anthocyanins and xanthones. Fruit colour is a major criterion used to judge maturity and also for grading of mangosteen fruit. During fruit development the fruit has at least six colour developmental stages. They are pale yellow green, blotchy pink, pinkish red, maroon red, dark maroon violet and violet black. Fruits are usually harvested at different stages according to colour, from light greenish yellow with scattered pink spots to dark purple. Mangoosteen is harvested during pinkish red stage for distant market while it is harvested at maroon red stage for nearby markets. After harvest, the purple colour continues to develop very quickly. For high fruit quality, the minimum harvest colour stage is that of distinct irregular, pink–red spots over the whole fruit. If fruits are harvested at light greenish yellow stage with scattered pink spots, the fruits do not ripen to full avour at all

3.1.19 Papaya

Fruits require 125 to 140 days from flowering to maturity. The change of colour from green to yellow and the consistency of the latex from milky to watery indicate that the fruit is ready for harvest. The mature fruit easily gets detached from stem either by turning it upwards or by twisting. There is change of skin colour from dark-green to light-green with some yellow at the blossom end (colour break). Portion of fruit exposed to sunlight becomes dark yellow in colour. Fruits are usually harvested at colour break to ¼ yellow for export or at ½ to ¾ yellow for local markets. A minimum soluble solids of 11.5 per cent is required by the Hawaiian grade standards. Under Indian condition, TSS at harvest should be minimum of 6 per cent.

3.1.20 Pear

Pear is the only temperate fruit that must be harvested mature and then ripened off the tree. Soluble solids, days from full bloom, iodine starch test, flesh firmness, ground colour, and ease of separation from the branch are some of the important harvest indicators. Soluble solids are not usually used as an indicator but it should be at least 13 °B for marketability. Days from full bloom provide idea of when the pears will be ready for harvest and then other methods can be used to pinpoint the best harvest date. The starch iodine test is in which the flesh of halved pears is dipped in an iodine solution and the per cent of area turning blue-black is estimated. The accepted amount of blue-black colour development for starch iodine test is 60 per cent or less for mature fruit. Pear firmness, determined with a pressure tester, should be 23 lb, but can be less if the soluble solids are less than 13 °Brix. Surface colour gradually changes from green to yellowish green. The fruit should come off the spur easily without tearing or breaking the fruiting spur.

3.1.21 Pineapple

Pinapple fruit is harvested on the basis of size, colour, or both, and twisted away from the stalk when mature. Shell colour changes from green to yellow at the base of the fruit. Pineapple is a non-climacteric fruit and should be harvested when ready to eat. A minimum soluble solids content of 12 per cent and a maximum acidity of 1 per cent assures minimum flavour acceptability by most consumers. It requires 105 to 130 days for maturity. Eyes on the fruit become smooth or flat and they slightly bulge out. Besides, eye bracts become loose and turn brown in colour with the right maturity. Mature pineapple should have 12 per cent T.S.S. 0.5 to 0.6 per cent acidity, with the specific gravity of 0.98 to 1.02.

3.1.22 Plum

Colour is the major harvest indicator for plums. Since each variety has its own characteristic colour change, familiarity with standards for planted varieties are important. Soluble solids should be at least 17 °Brix and pressure testing may be useful.

3.1.23 Pomegranate

Pomegranate fruits ripen in about 5-6 months from the time of flowering. Mature fruits should be immediately removed from the plants as delay in harvesting leads to fruit cracking. Fruits harvested at a premature stage have poor keeping quality and are prone to damage during handling and transport. The fruits are harvested when outer rind becomes yellowish and the fruit when tapped produces a metallic sound. Complete drying of floral parts could also be considered as an index for maturity. The round fruits become flattened from all the sides and basal beak shaped portion shrinks at the time of maturity. In general, acidity of juice should be below 1.85 per cent. Depending upon the variety, fruits are harvested when the TSS/acid ratio reaches 32-37. Bhagwa and Ganesh the most popular varieties are harvested respectively 180 and 150 days after anthesis with respective TSS/acid ratio of 32 and 37.

3.1.24 Sapota

The erratic flowering habit of sapota and the presence of fruit at all stages of development on the tree make it difficult to determine optimum harvest time (Lakshminarayana, 1980). Fruits harvested later than optimum time usually soften very rapidly and become very difficult to handle. Fruits harvested earlier than physiological maturity may not soften, are usually low in sweetness and high in astringency when ripe, with a rather unappealing alcoholic aftertaste, and form pockets of coagulated latex that lower quality. Unripe fruits are highly astringent and contain large amounts of leucoanthocyanidins. The sucrose content and pulp-to-peel ratio increase during maturation. The fruits shed off brown scaly external material and become smooth when reaching physiological maturity. Fruit ready for harvest do not show a green tissue or latex when scratched with a fingernail. Fully mature fruits have brown skin and separate easily from the stem without leaking latex. Extent of scruffiness is also a good indicator of maturity. A fruit with a smooth surface, shining potato colour and rounded styler end is considered mature. Fruits generally mature in about 240 to 270 days after flowering. Brown scaly material disappears from the fruit surface as the fruits approach full maturity. The dried spine like style at the top of the fruit falls or drops off easily when touched.

3.1.25 Strawberry

Strawberries owing to their non-climacteric nature, can be harvested when two-thirds of the fruit is red in colour. Fruits should be handled very carefully, by not holding more than a few berries in hand at a time and use shallow containers so that these are not bruised by the weight of fruits piled on top. The TSS acid ratio varies between 5.35 and 9.28 with 25 per cent and 100 per cent colour, respectively. Change in colour is the most common indicator used for harvesting strawberries.

3.1.26 Tamarind

Tamarind fruits take about 8 months from fruit set to harvest. As pods mature, skin develops into a brown, brittle shell, the pulp turns brown or reddish-brown, and seeds become covered with dry and sticky pulp. When fully ripe, the shells are brittle and easily broken. Mature fruits can be left on the tree for more than 6 months after ripening without significant spoilage, however birds and insects become pests. Fruit should be harvested when the moisture content is < 20 per cent to facilitate separation of the shell from the pulp.

3.2 Vegetables

3.2.1 Beet

Size is the harvest indicator for beet roots. Most beets are harvested when they are 1 to 3 inches in diameter. Baby size beets can be harvested at a smaller size. Beets prepared with tops are also harvested at a smaller size when the tops are still tender. Standard size beets can be sold with the tops intact, topped with stems no more than 1D2 inch long or short-trimmed with stems no more than 4 inches long.

3.2.2 Bell Pepper

Generally bell peppers are harvested in the mature green stage. These should be of good shape, waxy, firm and shiny. Peppers for canning are allowed to attain full size and develop a red colour before harvesting.

3.2.3 Brinjal

The fruits should be harvested when they are still tender but after they have attained a good size and when the surface is bright and glossy. At harvesting the calyx end the stem end are kept attached to the fruit. The fruits become ready for first picking in about 120-130 days of seed sowing depending on the variety. Fruits are harvested when they become greenish yellow or bronze and their flesh turns dry and tough. Pressing the thumb against the side of the fruit can indicate the maturity of the fruit. If the pressed portion springs back to its original shape, the fruit is too immature.

3.2.4 Broccoli

Heads should be harvested when the florets are tight and the colour is rich green with a purplish cast. Terminal head diameter should be at least 7.5 cm. Smaller heads can be bunched to achieve the specified count for the carton. Stems and heads should be cut at about 20 cm in length. Side shoots and large leaves should be trimmed leaving about 15 cm bare stem. Heads which are yellowish in colour and with open florets should be discarded.

3.2.5 Brussel's Sprouts

Harvesting after a moderate freeze will improve the quality of Brussel's sprouts immensely as these become sweeter and more palatable. Harvesting from the bottom of the stem up and harvesting only firm sprouts that are 1 to 2 inches in diameter is recommended. Outer leaves should be removed.

3.2.6 Cabbage

Maturity is indicated by compactness of the head which can be only measured by slightly compressing it with moderate pressure. Delaying harvest can result in split heads and increased incidence of field disease. Harvesting immature heads, however, reduces yield. They have tendency to burst or loosen the leaves beyond the marketable stage. At low elevations cabbage varieties mature in about 65 to 110 days from field setting while at high elevations it would take about 80 to 125 days. Solidity and firmness of head are the usual maturity characteristics used to judge the maturity. Head colour turns a lighter shade of green when full development is attained.

3.2.7 Cauliflower

The best stage of maturity is determined by curd size and condition of cauliflower. Local growers usually harvest the curd of desired size and before the curds become discoloured, loose or otherwise blemished. Mature curds are at least 15 cm in diameter. Loose floral parts are signs of over maturity. Cauliflower is ready for harvest at 90-120 days after planting. Depending upon the variety the curds should be harvested promptly when they are of full size but still compact, white and smooth. In over mature head flower stalks elongates resulting in loose, leafy condition.

3.2.8 Chilli

The stage of maturity at which chillies are picked depends on the type and purpose for which they are grown. For instance, these are picked either green for pickle making or fully ripe for drying. while they are picked at still green but full grown stage for vegetable purpose.

3.2.9 Cucumber

Cucumber as vegetable is ready for harvest within 45 to 55 days after sowing. The proper stage of maturity is judged by size and not by the age of the fruit. Cucumber for slicing should be picked when these are 15 to 35 cm long, whereas for pickling when 6 to 15 cm long. In general cucumber may be picked at any stage of fruit growth, provided yellowish colour should not develop

3.2.10 Drumstick

Drumstick (*Moringa oleifera*) plantation raised from limb cutting come to bearing within one year. Normally size is the harvest indicator for moringa which takes 30-45 days for maturity from flowering. The maturity index for moringa varies with the region and purpose. In eastern India, very young and soft seeded pods are preferred while in south, people prefer somewhat hard and thick pods.

3.2.11 French Bean

French beans are harvested at different developmental stages depending upon the purpose. The vegetable beans are harvested 90 days after sowing where as the seed beans are harvested 120-130 days after sowing. Vegetable beans must be picked up at the proper stage of maturity, because they start losing their marketability after reaching the edible stage. At maturity, the colour of the pods changes from dark green to light green and they are well filled up. Quality depends upon sugar content and tenderness. With increasing maturity and size, sugar content declines rapidly, with a correspondence increase in starch and protein. High sugar content is indication of high quality. Ease of snapping and days from flowering are the most commonly used indices to determine maturity.

3.2.12 Gourds

Proper stage of maturity in bottle gourd, sponge gourd and ridge gourd is judged by their fruit size. Easy penetration of fingernail on a gentle press on the skin is a rough indication of maturity. The fruit should possess fine tender flesh with immature tender seeds. In most cases fruits require 12 to 15 days after pollination to attain marketable size.

3.2.13 Knol-Khol

Harvesting of knol-khol is done when it attains its marketable size of about 5 to 7 cm in diameter and bright colour depending upon the variety. At this stage the edible portion is tender and non-fibrous.

3.2.14 Muskmelon

Fruits are ready for picking in about 110 days depending upon variety. Fruits begin to separate from stem when these are mature. Muskmelon maturity is classified into two stages *viz.* Full slip and half slip stage. In full slip stage the fruit when mature slips out easily from the vine with little pressure or jerk. In half-slip stage fruits require 1-2 days more for reaching the full slip stage. Fruits for long distant transportation are harvested at half-slip stage. Apart from this development of netting on the surface is also another indicator of maturity of certain muskmelon cultivars. Internal colour, total soluble solids and firmness are some other maturity indices used for musk melon. In some Indian cultivars, green stripes on the skin begin to turn yellow during maturity. Formation of abscission zone, slips and netting are most common maturity indices for musk melon (Figure 9).

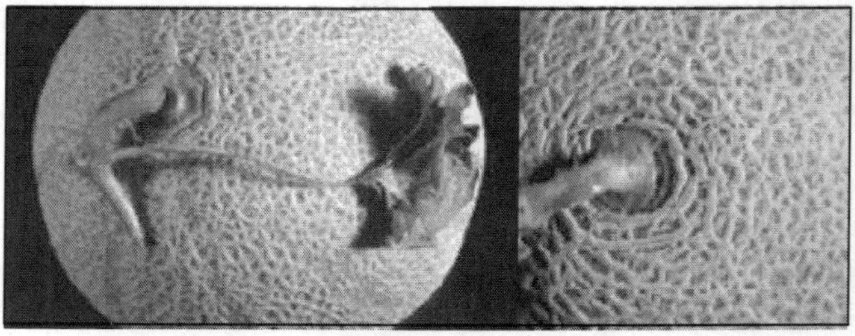

Figure 9: Maturity of Muskmelon

3.2.15 Okra

Young green fruits of okra are ready for harvest in about 45-60 days after seed sowing depending upon variety and season. Okra should be harvested when the fruit is bright green, the pod is fleshy with small and tender seeds. Pods should be selected which are 4 to 9 cm in length, green, tender and readily snap at the tip. They are harvested at an interval of 2 to 3 days, best time of harvest being 6 to 7 days after opening of flower. Mature pods are fibrous, tough, and unfit for human consumption.

3.2.16 Pea

Peas must be picked up at proper stage of maturity, because they start losing their sweetness rapidly after reaching the edible stage. The early varieties sown in October are ready to be handpicked in 50 days after planting while the mid season varieties sown in November are ready in 80-85 days. At maturity, the colour of the pods changes from dark green to light green and they are well filled up. Quality depends upon sugar content and tenderness. With increasing maturity and size, sugar content declines rapidly, with a corresponding increase in starch and protein. High sugar content is indication of high quality.

3.2.17 Pumpkin

Fruits require 110 to 120 days from sowing to maturity. Fruits should be harvested at full stage of maturity when the rind has hardened and its colour changes from green to yellow or orange though some varieties remain green even after full maturity.

3.2.18 Tomato

At any given time, six different maturity stages are generally found during fruiting of tomato. The stage of maturity for harvesting in tomato depends upon the purpose for which they are used. For long distance transportation; the fruits are harvested when they are firm, mature and greenish yellow or breaker stage. They get fully ripe or nearly red in 75-90 days after transplanting (Figure 10).

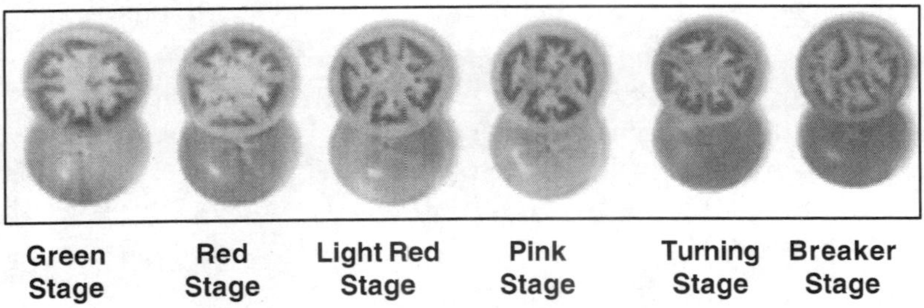

Figure 10: Maturity Stages of Tomato
(Source: http://www.tomato.org/)

a) Mature Green Stage

Fruits to be shipped are harvested when dark green colour changes to a reddish pink shade A simple way to determine maturity is to slice the tomato with a sharp knife. If seeds are cut, the fruit is too immature for harvest and will not ripen properly.

b) Turning Stage

Fruits develop pink or red colour on over 10 per cent but not more than 30 per cent of the tomato surface.

c) Breaker Stage

Dim pink colour observed on ¼ part of the fruit. Fruits are harvested at this stage to ensure the best quality. Such fruits are less prone to damage during shipment and often fetch a higher price than less mature tomatoes.

d) Pink Stage

Fruits are stiff and nearly whole fruit turns reddish pink. Fruits for local sale are harvested at this stage.

e) Light Red

Pinkish-red or red colour shows on over 60 per cent but red colour covers not more than 90 per cent of the tomato surface

f) Red Stage

Fully ripened fruits are soft having dark red colour. Such fruits are used for processing.

3.2.19 Water Melon

Watermelons are harvested at full maturity as they typically neither develop internal colour nor increase sugar content after being removed from the vine. The ground spot (portion of the melon resting on soil) changes from pale white to a creamy yellow at proper maturity. However, size of the fruit and colour of the rind are not good indicators. Fruits are ready for harvesting in 90 to 120 days from sowing. Other common indicators for identifying maturity of water melon are withering of tendrils accompanying the fruit and production of heavy dull sound on thumping.

Destructive sampling is used to judge maturity of a population of watermelons. For seeded cultivars, maturity is reached when the gelatinous covering (aril) around the seed is no longer apparent and the seed coat is hard. Cultivars vary widely in soluble solids at maturity. In general, a soluble solid content of at least 10 per cent in the flesh near the center of the melon is an indicator of proper maturity. Flesh should also be firm, crisp and of good colour.

3.3 Tuber, Root and Bulbous Crops

3.3.1 Arrow Root

The crop attains maturity in 10-11 months after planting. Maturity is indicated by yellowing, wilting and drying up of the leaves. At this stage, the plants are dug-out and rhizomes are separated from the stem.

3.3.2 Carrot

Typically carrots are harvested at an immature state when the roots have achieved sufficient size to fill in the tip and develop a uniform taper. Depending upon variety carrot becomes ready for harvesting within 100 to 120 days from sowing. At the time of harvest the roots should not be pithy or solid. Depending on the variety, the desired size is the primary consideration in harvesting the roots. At marketable stage carrots should have at least 2.5 to 4 cm diameter at the upper end.

3.3.3 Elephant Foot Yam

Harvesting of large tuberous corm of elephant foot yam is done in 8-10months at senescence stage. Corms are harvested by digging when green part above soil is completely withered.

3.3.4 Garlic

Garlic is ready for harvesting when the tops turn yellowish or brownish, show signs of drying up and bend over. Normally it is harvested when 50 per cent of the plants have shown neck fall. Crop gets ready for harvest in about 100 to 140 days after planting.

3.3.5 Onion

Harvesting depends upon the purpose for which the crop is planted and maturity will depend upon the variety being used. "Field-dry" maturity is indicated when bulb neck is completely dry to the touch and tops are fallen over. Best time to harvest *rabi* onion is one week after 50 per cent tops have fallen over while in *kharif* season, since tops do not fall, bulbs are harvested soon after the colour of leaves change to light yellow and bulbs develop red pigmentation on them. Onion crop is ready for harvesting in five months for dry onion. However, for marketing as green onion, the crop becomes ready in three months after sowing. Green onion is best when they are like lead in pencil thickness and until a small bulb is formed. Development of red pigment and the characteristic pungency of the variety are also important harvest indices.

3.3.6 Potato

Potatoes can be harvested for immediate consumption in an immature state, usually from the time plants reach full flowering stage. At this time the skin of the tuber remains thin and soft but these potatoes are not suitable for storage. Potatoes harvested for bulk storage should not be harvested until at least two weeks after the plant tops have dried off, by which time the tubers are fully mature and covering skins mostly developed. They are then less susceptible to damage as well. Early varieties mature between 80 to 100 days, whereas late varieties are ready for harvest in 120- 160 days. Skins slipping from the tuber, starch content and leaf senescence or top drying are the main harvest indices used for potato.

3.3.7 Radish

Edible roots of radish mature between 30 to 70 days post emergence depending on variety. The temperate types reach harvest maturity 25-30 days after seed sowing while tropical varieties require longer period. At the time of harvest, the roots should not be pithy or solid. European types are ready for harvesting within 30 days from sowing while Asiatic types mature within 45 to 60 days from sowing. At this stage roots are mild, tender and crisp and usually of the proper marketable size. Roots must be harvested before they become pithy, bitter and fibrous.

3.3.8 Sweet Potato

Sweet potatoes are harvested 3.5 to 4 months after planting. Early varieties like Kalmegh or Samrat can be harvested in 90 days. Harvesting is done when the leaves turn pale and start yellowing. Moreover after cutting the tuber if white flesh remains white, then they are ready but if it turns blackish or green after exposure then tubers are said to be immature for harvest. In general, delay in harvest after 120 days increase sweet potato weevil infestation.

3.3.9 Tapioca

The tubers of cassava plants are ready for harvest in 7-11 months depending on the variety. Sree Prakash, H-119, CI-590, Sree Jaya and Sree Vijaya are short duration varieties maturing in 6-7 months. H-97 and H-165 takes 7-8 months; H-226, 8-9 months and Sree Sahya, Sree Visakam, Sree Rekha, Sree Prabha are long duration crops which take 9-10 months for maturity. When the soil shows cracks trail digging is done. The leafy portions of the plants are discarded before harvesting of roots.

3.3.10 Taro

Normally roots of taro are harvested after 200-210 days. Harvesting at 165 days after planting has been found ideal in states of Andhra Pradesh, West Bengal and Uttar Pradesh when grown as a summer crop. Yellowing of leaves is an indication of crop maturity when cultivated in dry condition.

4.0 Grade Standards

The Codex Alimentarius Commission was established by the FAO and WHO in 1963 for harmonization of International Food Standards, developing guidelines and codes of practice for protect the health of consumers and ensure fair practices in the food trade. The commission also promotes coordination of all food standards work undertaken by international governmental and non-governmental organizations. Increasing number of countries are now aligning their food standards with those of the Codex Alimentarius.

In India, APEDA took due cognizance of the above international developments and embarked on a programme to financially extend assistance to implement GAP, which is an internationally recognised quality assurance system for growing fruits and vegetables. Accordingly the quality specifications of vegetables for export are listed in Table 4.

Table 4: Quality Specifications of Vegetables for Export

Bitter gourd	Green, lustrous, more prickly 20-25cms long with short neck
Bottle gourd	Green, tender, cylindrical, 20-25cms with soft hairs
Brinjal	Round, thorn less, dark green calyx, purple with good lustre
Chillies	Green,lustrous, 7-8cms, smooth without wrinkles
Cluster bean	Green, tender, 7-10cm in length
French bean	Tender green, 15 cms in length
Garlic	White round, 3.5-4cms having 15-20 big sized cloves
Okra	Green, tender, 6-8cms, 4-5 ridges, slender and roundish with pointed tips
Onion (Big)	Light – dark red, globose, 4-6cms in size and strongly pungent(Bangladesh, SEA, Gulf) yellow or brown for Europe and Japan; TSS: 8-10 per cent
Onion (Multiplier)	Bigger cloved bulb of bright red colour
Onion (Small)	2-3cm dark red and round shape
Potato	White skin, long and extra long tubers with yellow- light yellow flesh(Iraq, Iran); white and red types oval shapes for Bangladesh
Tomato	Round, medium sized, red colour for Middle East and cherry tomatoes for Europe,TSS acid ratio – 10: 1

The grade and standards for different crops are as under.

4.1 Fruit Crops

The general minimum requirements for all the fruit crops as per codex standards are mentioned below. In all classes, subject to the special provisions for each class and the tolerances allowed, the produce must be sound, should not be affected by rotting or deterioration such as to make it unfit for consumption. The produce should be clean, practically free of any visible foreign matter; practically free of pests and damage caused by them affecting the general appearance of the produce; free of abnormal external moisture, excluding condensation following removal from cold storage; free of any foreign smell and/or taste; practically free of damage caused by low and/or high temperatures. In addition, the produce must be: whole; well formed; normally developed.

4.1.1 Apple

a) EU Standards

i) **Extra class**: Apples in this class must be of superior quality. The flesh must be sound. They must be characteristic of the variety. They must be free of defects, with the exception of very slight superficial defects, provided these do not affect the general appearance of the produce, quality and presentation. Five percent by number or weight of apples not satisfying the requirements of this class but meeting the requirement of class-I is allowed. Not more than 1.0 per cent for apples affected by decay or internal breakdown is allowed in this class.

ii) **Class I** : Apples in this class must be of good quality. The flesh must be sound. They must be characteristic of the variety. Slight defects may be allowed, provided these do not affect the general appearance of the produce quality, keeping quality and presentation. Ten percent by number or weight of apples not satisfying the requirements of this class but meeting other the requirement of class-I is allowed. Not more than 1.0 per cent for apples affected by decay or internal breakdown is allowed in this class.

iii) **Class- II:** This class includes apples which do not qualify for inclusion in the higher classes, but satisfy the minimum requirements specified above. The following defects, however, may be allowed, provided the apples retain their essential characteristics as regards the quality, keeping quality, presentation, defects in shape and development; defects in colouring; skin or other defects.

b) Indian Standards

i) **Super large :** 80 mm and above in diameter

ii) **Fancy small :** 60 –70 mm in diameter

iii) **Pittoo :** 55-60 mm in diameter

iv) 5-10 mm blemishes and 10-20 per cent russetting is allowed

4.1.2 Banana

In all classes, subject to the special provisions for each class and the tolerances allowed, the bananas must be: - whole (taking the finger as the reference); - sound, produce affected by rotting or deterioration such as to make it unfit for consumption is excluded; - clean, practically free from any visible foreign matter; - practically free of pests affecting the general appearance of the produce; - practically free of damage caused by pests; - free of abnormal external moisture, excluding condensation following removal from cold storage, and bananas packed under modified atmosphere conditions; - free of any foreign smell and/or taste; - firm; - free of damage caused by low temperatures; - practically free of bruises; - free of malformation or abnormal curvature of the fingers; - with pistils removed; - with the stalk intact, without bending, fungal damage or desiccation.

a) EU Standards

i) **Extra class**: superficial blemishes not covering more than 1 sq cm, tolerance of 5 per cent by weight.

ii) **Class I** : superficial blemishes not covering more than 2 sq cm, tolerance of 10 per cent by weight.

iii) **Class II** : superficial blemishes not covering more than 4 sq cm.

iv) Length : 14-27 cm.

4.1.3 Citrus

a) EU Standards

i) Extra class: tolerance level : upto 5 per cent

ii) Class I : upto 10 per cent

iii) Class II : upto 10 per cent

iv) Class III : upto 15 per cent

b) Indian Standards

i) Extra special, Special, Good, A, B, C.

ii) Min diameter: 70-80 mm for Malta, 65-75 for mosambi and 70-85 for loose mandarins.

iii) 3-5 per cent tolerance allowed.

4.1.4 Grape

a) EU Standards

i) **Extra** – minimum weight: 200g, 5 per cent tolerance by weight is allowed, packed in a single layer.

ii) **Class I** – minimum weight: 100-150g, 10 per cent tolerance by weight.

iii) **Class II** – minimum weight: 75-100g.

iv) **Class III** – minimum weight: 75g, tolerance level of 15 per cent by quality and size.

b) Indian Standards

i) **Class Special**: Under this category bunches should be compact, uniform size, two-thirds of berries should comprise of 20 mm in diameter and pale green colour. The minimum accepted length under this category is 250 mm having weight of 700g.

ii) **Class-A:** Under this category bunches should be compact, uniform size, two-thirds of berries should comprise of 15 mm in diameter and pale green colour. The minimum accepted length under this category is 200 mm having weight of 500g.

ii) **Class-B**: Same as A with minimum length: 150 mm and weight: 380g.

4.1.5 Mango

In addition to general standard the peduncle shall be no longer than 1.0 cm as per codex requirement.

a) EU Standards

i) **Extra class**: Under this class, each fruit should have minimum weight of 200g and free from defects. Five percent tolerance by number or weight is allowed in this class.

ii) **Class I**: Under this class, fruits may have slight defect due to sap exudation and small healed bruises. The tolerance limit in this class is 10 per cent.

iii) **Class II**: Under this class, the fruits may have larger and healed bruises.

b) Indian Standards

The grades vary with variety. For Alphonso and Kesar, the commercially accepted grades are as under:

i) A+ grade: >300 g

ii) A grade: 250-299g

iii) B grade: 200-249 g

iv) C grade: 150-199 g

v) D grade <150

4.1.6 Plum

a) EU Standards

i) **Extra class** : 30-35 mm, tolerance of 5 per cent by weight

ii) **Class I** : 25-28 mm, 10 per cent tolerance

iii) **Class II** : less than 25 mm, 10 per cent tolerance

iv) Not more than 2 per cent plums in Class I and 4 per cent in Class II should be split or worm eaten.

b) Indian Standards

i) **Special :** min dia of 41mm

ii) **Grade I :** 35 mm

iii) **Grade II** : 29 mm

iv) Should be uniform greenish-yellow to yellow colour

4.2 Vegetable Crops

4.2.1 Beans

a) EU Standards

i) Extra Class : tender, seedless and stringless, tolerance of 5 per cent by weight, 6 mm in size.

ii) Class I : Slight discolouration, small seeds, short and soft strings, tolerance of 10 per cent by weight.

iii) Class II : Beans which do not qualify for Class I.

4.2.2 Brinjal

a) EU Standards

i) **Class I :** cracks covering not more than 3 sq cm, tolerance of 10 per cent by weight.

ii) **Class II** : cracks covering not more than 4 sq cm, tolerance of 10 per cent by weight.

iii) **Class III** : may be slightly fibrous and show considerable seed development, cracks covering not more than 6 sq cm, tolerance of 15 per cent by weight.

iv) The diameter 40 mm for elongated brinjal and 70 mm for globus brinjal.

4.2.3 Carrot

a) EU Standards

i) **Extra Class :** green and purple tops are not allowed, tolerance of 5 per cent by weight

ii) **Class I** : green or purple tops upto 1cm long, tolerance of 10 per cent by weight

iii) **Class II** : green or purple tops upto 3cm long, tolerance of 10 per cent by weight, broken tops not more than 25 per cent

iv) Early carrot roots must be 10-40mm in diameter and 8-150g in size

v) For storage carrot roots must be 20-45mm in diameterand 50-200g in size

4.2.4 Garlic

a) EU Standards

i) **Extra class** : compact cloves, roots must be cut-off, tolerance of 5 per cent by weight, 45 mm in dia.

ii) **Class I** : tolerance of 10 per cent by weight, not more than 1 per cent should have visible sprouts, 30 mm in dia.

iii) **Class II** : tolerance of 10 per cent by weight, not more than 5 per cent should have visible sprouts, 30 mm in dia.

b) Indian Standards

i) Big : 30mm

ii) Medium : 25mm

iii) Small : 10mm

iv) Mix : 10-12mm

v) Tolerance of 5 per cent by weight

c) U.S. No. 1.

"U.S. No. 1" consists of garlic of similar varietal characteristics which is mature and well cured, compact, with cloves well filled and fairly plump, free from mold, decay, shattered cloves, and from damage caused by dirt or staining, sunburn, sunscald, cuts, sprouts, tops, roots, disease, insects, or mechanical or other means. Each bulb shall be fairly well enclosed in its outer sheath. Unless otherwise specified, the minimum diameter of each bulb shall be not less than 1-1/2 inches.

Tolerances. In order to allow for variations incident to proper grading and handling, the following tolerances, by weight, are provided as specified:

- **For defects.** Ten percent for garlic in any lot which fails to meet the requirements of this grade, including therein not more than 2 percent for garlic which is affected by decay.
- **For size**. Five per cent for garlic in any lot which fails to meet any specified size.

d) Unclassified

"Unclassified" consists of garlic which has not been classified in accordance with the foregoing grade. The term "unclassified" is not a grade within the meaning of these standards but is provided as a designation to show that no definite grade has been applied to the lot.

4.2.5 Onion

a) EU Standards

i) Class I : blemishes should not cover 10 per cent

ii) Class II : blemishes should not cover 10 per cent

iii) Class III : blemishes should not cover 20 per cent

b) Indian Standards

i) Extra big : min dia : 60mm

ii) Big : 30-40mm

iii) Medium : 20-30mm

iv) Small : 15-20mm

v) Mix : 15-20mm

vi) Not more than 2 per cent should be decayed, tolerance of 5 per cent by weight

4.2.6 Pea

a) EU Standards

i) Class I : well filled containing at least 5 seeds, tolerance of 10 per cent by weight

ii) Class II : 3 seeds

b) Indian Standards

i) The tolerances for the standards are on a container basis. However, individual containers in any lot may vary from the specified tolerances as stated below, provided the averages for the entire lot, based on sample inspection, are within the tolerances specified.

ii) For a tolerance of 10 percent, individual containers in any lot may have not more than one and one-half times the tolerance.

iii) For a tolerance of less than 10 percent, individual containers in any lot may have not more than double the tolerance.

c) Unclassified

Consist of peas which have not been classified in accordance with the foregoing grades. The term "unclassified" is not a grade within the meaning of these standards but is provided as a designation to show that no definite grade has been applied to the lot.

d) US Standards

U.S. No. 1. Consist of pods of peas of similar varietal characteristics which are not overmature or excessively small, not badly misshapen or watersoaked, and which are fairly well filled, fresh, firm, free from decay, and from damage caused by black calyxes, freezing,splitting, hail, dirt, leaves, or other foreign matter, mildew, or other diseases, insects, or mechanical or other means. The peas shall be at least fairly tender, free from decay, and from damage caused by split skins, disease, insects or mechanical or other means.

Peas may be specified as "U.S. No. 1 Green Calyxes" when they meet the requirements of U.S. No. 1 grade and an average of three-fourths or more, by weight, of the pods in any lot, but not less than one-half of the pods in each container, have calyxes which are of a fairly good green colour.

U.S. Fancy. U.S. Fancy shall consist of pods of peas which are well filled and have an average of three-fourths or more by weight, of the pods in any lot, but not less than one-half of the pods in each container with calyxes which are of a fairly good green colour, and which meet the requirements of U.S. No. 1 grade in all other respects.

5.0 Conclusion

Quality determination of fruits and vegetables is at present characterized by the overwhelming use of sensory evaluation. Quality indices include primarily external properties, such as appearance, size, shape, colour, texture, and flavour and labeling controls. Conformity with relevant product standards is also an important requirement that is inspected for the trade of fruits and vegetables. Maturity measures made by the producers, handlers and quality control personnel must be simple, readily performed in the orchard and require relatively low operational cost. The index should be objective (quantitative) rather that subjective (descriptive). The developed index should be nondestructive to avoid loss of produces. The index must be consistently related to the quality and postharvest life of the commodity for all growers and for all through the year. With the increasing competitive and concentrated markets, more important becomes the introduction of systems and technologies to manufacture produce differentiated by their presentation for positioning in these markets. The production of high quality fruits and vegetables and the maintenance and enhancement of this quality in postharvest and distribution operations is associated with careful incorporation of technologies applied throughout the production, harvesting and postharvesting stages. These technologies are crucial to ensure quality and safety. The principle is not to use sophisticated technologies, but to handle efficiently the produce throughout the chain. Only in this way will quality and safety be maintained and postharvest losses can be reduced. A new approach to describe fruit and vegetable quality by an overall evaluation method provides numerical values that could be used for their classification in the future.

6.0 Important References

Behera, T.K., Pal R.K., Sen Nita and Singh Manoj 2004. Effect of maturity at harvest on physico-chemical attributes of sweet pepper (*Capsicum annum* var grossum) varieties. *Ind. J. Agril Sci.*, **74(5):** 251-3.

Deodhar, S.Y. and Dave, H. 2001. Managing Quality of Agri-commodities and Food Products under WTO Regime. **In.** *Implications of WTO Agreements for Indian Agriculture,* S. Datta and S. Deodhar (Eds.). Oxford-IBH, Delhi.

Dhatt, A.S., Mahajan, B.V.C., Sandhu, K.S., Garg, A. and Sharma, S.R. 2007. *Handbook on Postharvest Handling of Fruits and Vegetables.* 3rd edn, PHPTC, PAU, Ludhiana.

FAO 2001. *Food Quality and International Trade*. http://www.fao.org/ag/ags/Agsi/gtz_html.htm.

Goel, A.K., Kumar, R. and Mann, S.S. 2007. *Postharvest management and value addition.* Daya publishing house, New Delhi-35.http://agmarknet.nic.in/fveggmrules04.htm#rules.

Jadhav, V.T. and Sharma, J. 2007. Pomegranate cultivation is very promising. *Indian Horticulture*, pp. 30-31.

Kader, A.A. 2002. *Postharvest Technology of Horticultural Crops.* 3rd Edn, University of California, Agricultural and Natural Resources, Publ. 3311, 296 p.

Kader, A.A. and Arpaia, M.L. 1992. Postharvest handling systems: subtropical fruits.

In: *Postharvest Technology of Horticultural Crops*, A.A. Kader (Ed.), Univ. of Calif., Davis CA, pp. 233-240.

Kitinoja, L. and Gorny, J. 1998. *Postharvest technology for fruits and vegetables Produce marketers: Economic opportunities. Quality and Food Safety*, Department of Pomology, University of California, Davis. A joint publication of UC Postharvest Outreach Program and Punjab Horticultural Postharvest technology Centre, USAID/ACE

Kute, L.S. and Shete, M.B. 1995. Sapota (Sapodilla). **In:** *Handbook of Fruit Science and Technology*, D.K. Salunkhe and S.S. Kadam (Eds.), Marcel Dekker Inc., NY, pp. 475-484.

Lakshminarayana, S. 1980. Sapodilla and prickly pear. **In:** *Tropical and Subtropical Fruits. Composition, Properties and Uses,* S. Nagy and P.E. Shaw (Eds.), AVI Pub, Westport CT, pp. 415-441.

Massantini, R., Carlini, P. and Anelli, G. 1997. NIR Spectroscopy to Evaluate Fruit and Vegetable quality During the Postharvest. *Industrie Alimentari,* **36:** (357) 321–326.

Pal, R.K. 2008. *Final Report of Consultancy Services on Controlled Atmosphere Storage of Fruits and Vegetables* submitted to M/S Ministry of Railways, Govt of India, July 2008, 40 p.

Petracek, P.D., Dou, H. and Pao, S. 1998. The influence of applied waxes on postharvest physiological behavior and pitting of grapefruit, *Postharvest Biol. Technol.*, **14**: 99-106.

Ryall, A.L. and Pentzer, W.T. 1974. *Handling, Transportation and Storage of Fruit and Vegetables*: Volume 2, Fruits and Tree Nuts, The AVI Publishing Company Inc.,Westport, CN.

Sargent, S. 1998. *Handling Florida vegetables - Eggplant*. http://hammock.ifas.ufl.edu.

Thompson, A.K. 1996. *Postharvest Technology of Fruits and Vegetables*. 1st Edn, Blackwell Science Inc., USA.Watada, A.E., Herner, R.C., Kader, A.A., Romani, R.J. and Staby, G. L. 1984. Terminology for the description of developmental stages of horticultural crops. *Hort Science*, **19**: 20-21.

Wills, R., McGlasson, B., Graham, D. and Joyce, D. 1998. *Postharvest: An Introduction to the Physiology and Handling of Fruit, Vegetables and Ornamentals*. 4th Edn, CAB International, UK.

Websites

http://www.postharvest.ucdavis.edu
http://www.herbs.ph/attachments/article/1610/
http://www.extension.udel.edu
http://www.cuke.hort.ncsu.edu
http://www.extension.missouri.edu
http://www.www.sakatavegetables.com

Chapter 4

Mechanisation in Harvesting

S.C. Mandhar

Emeritus Scientist, ICAR-Indian Institute of Horticultural Research, Hessaraghatta, Bengaluru – 560 089, Karnataka, India
E-mail: suresh.mandhar@gmail.com

R.T. Patil

Former Director, ICAR-Central Institute of Postharvest Engineering & Technology, P.O. PAU, Ludhiana – 141 004, Punjab, India
E-mail: ramabhau@gmail.com

1.0 Introduction

Harvesting of fruit from the trees is a cumbersome and time consuming operation. The fruits are generally required to be plucked individually by holding the fruit at their natural break point firmly but gently. Sometimes due to shortage of labour fruits are harvested by shaking the tree manually as is done for mango by many farmers. The fruits fall on the ground and then picked up. This injures the fruit internally and the spoilage increases during ripening. The mango fruits are also harvested manually by using conventional harvesting aids like 1) bamboo stick folded to form elliptical shape frame and tied on one end of a bamboo or any light wooden pole. The hand woven cotton net is tied below the frame to collect the fruits, 2) some conventional harvesters are fitted with a cutting notch to harvest the fruits with pedicel and a canvas collection sack to collect the fruits, 3) bamboo pole fixed with a semicircular or sickle blade and net at the distant end and a net to collect the harvested fruits. The conventional mango harvesters have been modified at Indian Institute of Horticultural Research, Bangalore, Central Institute for Subtropical Horticulture, Lucknow and Dr. Balasaheb Sawant Konkan Krishi Vidyapeeth, Dapoli, to increase the harvesting capacity and increase the shelf life of fruits by harvesting these along with 2 to 3 cm. long pedicle.

Timely harvesting of fruits is important for maintaining the transportability, quality and shelf life. However, the harvesting capacity of these harvesters is very low as the individual fruits are to be harvested and collected. Hence there has been a need to develop high capacity machines for this purpose. For adopting the mechanical harvesters the tree canopy has to be properly trained. Some of the machines which have been developed and are being used in different countries are presented in this chapter.

2.0 Harvesting Aids

Harvesting aids have been developed to increase the harvesting capacity of the person and to reduce the damage to the fruits. Generally one person is required to use these aids to harvest the fruits. The harvesting capacity depends upon the yield of fruit tree and type of fruit. The Mango harvester developed at Indian Institute of Horticultural Reaearch, Bangalore (Figure 1) is an improvement on the conventional harvester and harvests the fruit with pedicel. The fruit is lowered in net in the opening in between frame and aluminium pole. While pulling the harvester, the fruit is held between the cushioned frame and the pole and the blade cuts the pedicel of 1-2cm length to harvest the fruit. There is no sap bleeding, fruits remain shining without blemishes and the shelf life of the fruit is increased by 3-4 days. The harvesting capacity of the harvester is 50 to 150 kg/hr depending upon the variety, yield and height of tree.

Figure 1: Mango Harvester

The sapota, guava and pomegranate harvester (Figure 2) have been developed at IIHR, Bangalore. It consists of harvesting frame, conical in shape and a pole. The harvesting frame is fitted with comb consisting of cushioned pegs to hold the fruit and harvest these while pulling the harvester. The fruits of required size only get harvested while the immature fruits are left on the tree. The capacity of sapota and guava harvesters is 50 – 100 kg/hr. In pomegranate harvester, the distance of pegs on the frame is more to suit the size of fruit to be harvested.

Figure 2: Sapota Harvester

Lime harvester (Figure 3) has been developed at IIHR, Bangalore. A small box of around 10cm dia and 20cm length is fitted with the hook. The harvester is pushed easily in the canopy, the fruit is held in the hook, harvested and collected in the box while pulling the harvester. The box is emptied after 8-10 fruits are collected in it. The lime harvester has the harvesting capacity of 15-20 kg/hr.

Figure 3: Lime Harvester

The cut and hold harvesters have secateurs head fitted at distal end. It is able to hold the fruit after harvesting (Figure 4). These are generally telescopic and the length can be extended to 3 meter.These are suitable for harvesting most fruits including mango and litchi. The pedicel of the fruit is cut and the fruit is lowered to the ground and put in the required container.

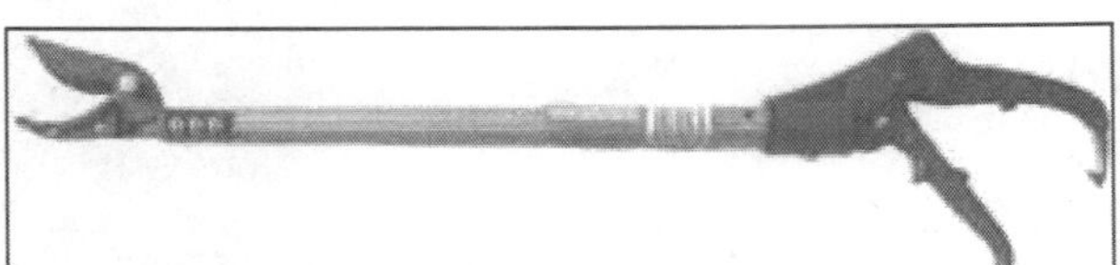

Figure 4: Cut and Hold Harvester/Pruner

2.1 Man Positoning System

Generally the manual labour climbs the tree to harvest the fruits in India. However this has to be avoided as there is a risk for the person to fall down. Portable aluminium light weight tripod ladder (Figure 5) can be used as a man positioning system to avoid the climbing of labours on the trees to harvest the fruits. Ladders are also very useful in pruning the tall trees. The double pole telescoping ladder and single pole ladder are suitable aids for harvesting coconut, areca nut, black pepper and betel wine.

A self propelled hydraulically operated man positioning system has recently been developed at CIAE Bhopal (Figure 6). A hydraulically operated three wheel self-propelled elevator system was designed and fabricated at CIAE Bhopal for harvesting of mango, oranges, oil palm etc. The unit can be rotated 360 ° and has hand-controls on platform, steering mechanism with a 11 hp petrol engine as prime mover. The

Figure 5: Tripod Ladder

Figure 6: Self Propelled Hydraulically Operated Man Positioning

platform could be elevated up to 6 meter from ground level, which makes it suitable for a variety of operations in medium height orchards including oil palm. Self propelled elevating platform can also be used for pruning, spraying and harvesting tall trees. One operator stands in the cage/platform and controls the height and movement of the platform. The fruits are plucked manually and put in the bag provided on the machine. The bag can be emptied after lowering the platform to the ground level. The hydraulic operated chain saw and pruners are also provided for pruning on these platforms.

In tractor operated mango harvester developed at Indian Institute of Horticultural Reserch, Bangalore labourers stand on the platform on the trailer (Figure 7) with harvesting aids for harvesting the individual fruits. The harvested fruits fall on the apron filled in crates/ bulk bins kept in the trailer.

Figure 7: Tractor Trailor Mounted Platform

2.2 Ultra Lightweight Harvesters

The light pole aluminum harvesters with pole length of up to 12 meter are available for harvesting coconut and oil palm (Figure 8). The aluminum pole is fitted with a sickle and the fruits are harvested by holding the bunch with the sickle and pulling the pole. The telescopic petrol engine (length 3.7 metre) operated vibrating sickle harvesters are available for harvesting the oil palm (Figure 9). The weight of petrol engine operated harvester is about 7 kg and the fruits can be harvested up to a height of 5 m. About 150 bunches can be harvested per day using this motorized harvester.

Figure 8: Light Pole Harvester for Oilpalm and Coconut

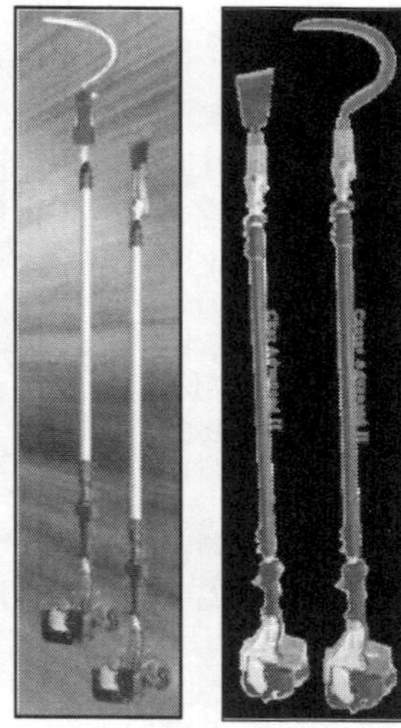

Figure 9: Engine Operated Harvesters for Oilpalm

2.3 Shaker Harvesters

The telescopic pole harvester with battery operated revolving reel is available and is suitable for harvesting olives (Figure 10a) The beater harvester for olives is operated by 26 cc/900 W petrol engine and produce 1150beats per minute (Figure 10b). The beater harvester has a harvesting capacity of 50-80 kg/hr. The shaker harvester for limb/branch is provided with 48.7 cc/2.2 kW petrol engine, weighs 13.9 kg and the pole length is 2.83 metre (Figures 10 c and d).

2.3.1 Apple

The shaker harvester for apple has been developed in USA (Figure 11). The canopy of the apple trees is trained such that shaker harvester can harvest it. The tree is generally trained into single or two branches (Y-shaped).

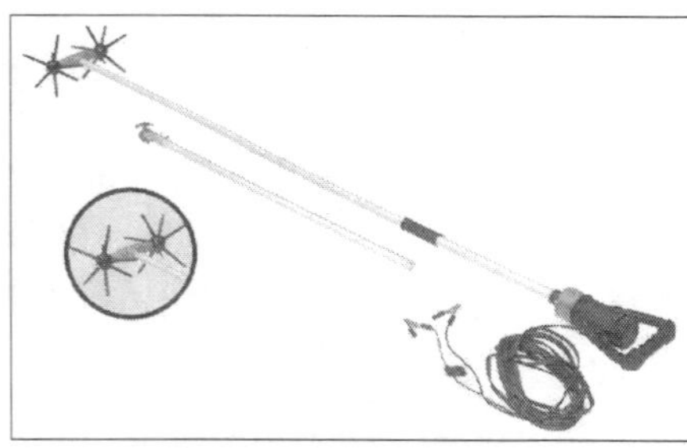

(a) Electrical shaker harvester

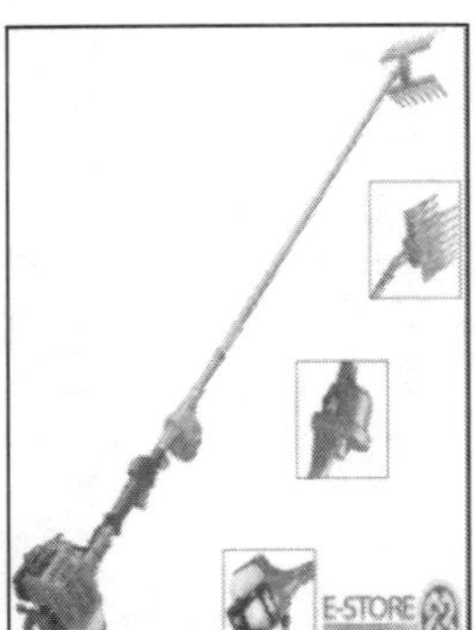

(b) Olive shaker harvester

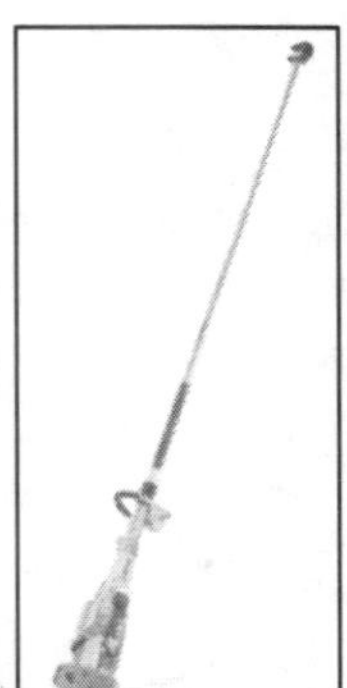

(c) Shaker harvester for branch

(d) Shaking olive branch

Figure 10: Shaker Harvesters

The individual branch of the tree is shaken by pneumatically operated cylinder by giving one jerk. The fruits get detached and fall on the collecting apron and are collected and conveyed to the truck using belt conveyors.

Figure 11: Shaker Harvester for Apple

2.3.2 Mango

The low cost vibrating harvester for mango developed at Asian Institute of Technology, Bangkok operates with 8.3 hp diesel engine. The individual branches/limbs of mango are held in padded clamps and vibrated with amplitude of 76-102 mm at 11 – 13 cycles per second. About 80-85 per cent fruits are removed. The fruits are collected on catching net supported on portable stands. Two to four trees can be harvested in an hour depending upon the tree size.

2.3.3 Nuts

The tractor operated harvester shakes the branches of the nut trees to harvest them (Figure 12). The stroke and frequency of the vibration can be adjusted. The fruits are collected by the catching frames.

2.3.4 Orange

Self propelled shaker harvesters (Figure 13) and tractor operated harvesters are available for shaking the trunk of the tree or branches of the tree. The stroke and frequency of the vibration can be adjusted. The fruits fall on the ground and then collected by fruit sweeps. The fruit pick up machines pick up oranges at 15-20 tonnes/ hr. at 97 per cent efficiency and caused 3 to 7 per cent split peels from handling. The fruits are collected by horizontal or vertical rotors, conveyed to the bins through conveyor belt and leaves and dust is blown away by blowers. The fruits are also collected by the catching frames.

a) Trunk Shake Catch Harvester for Orange

The trunk shake catch harvester (developed at Compton Enterprises Inc., Chico, California, U.S.A.) removed 90-94 per cent at harvesting capacity of 10 tonnes/hour. The orange tree is shaken for 5 to 10 seconds at a frequency of 8 to 10Hz. The shake – catch harvester had 2 sets of unbalanced masses totalled 125 kg had an eccentricity of 22 cm and rotated in same direction at 8 to 10 Hz (480 to 600 rpm).

2.4 Canopy Shaker Harvester

2.4.1 Blue Berries

The spiked drum shaker harvester has also been developed for harvesting blue berries (Figure 14). The fruits are collected on the catching aprons/conveyor belts below the harvester and conveyed to the storage bin through conveyor belts. The

Figure 12: Self Propelled Shaker Harvester with Inverted Umbrella Collection Apron and Belt Conveyor

Figure 13: Self Propelled Tree Shaker

fruits are also graded by the machine and sorted manually by 2-3 labourers. The harvesting capacity is around 400 – 500 kg/hour. The fruits are suitable for fresh consumption (eating)/table purpose. The harvester machine can harvest the blue berry grown on trellis. Around 80 per cent yield is marketable. The harvester with spiked tooth type drums for blue berry can be easily modified for harvesting Ber (*Ziziphus* sp.) and Aonla (*Phyllanthus emblica*).

Figure 14: Spiked Drum Shaker Harvester

2.4.2 Orange

Spiked drum type harvester (Figure 15) has been developed in USA and Australia to harvest the orange. The planting of the trees is done at specified spacing and the canopy is shaped. The tree is brought in between the spike drums of 6m dia. and the rotating drums are pushed inside the canopy, which shake the branches of the tree and harvest the orange. The fruits fall on the catching frame/net and from there these are conveyed to the truck by using belt conveyors. The harvested fruits are suitable for processing or juice extraction. The canopy harvester (developed by United States Department of Agriculture) has 2 shaker drums with eight sets of horizontal whorls, 3.2 cm diameter nylon spokes mounted on each of the two vertical shaker shafts (Figure 15). Each horizontal whorl set contained 16 spokes, 1.5 meter long beyond the shaft/hub to which they were bolted. The vertical spacing between horizontal whorls is 38 cm and top set is about

Figure 15: Canopy Shaker Harvester for Orange

3.7 meter above the ground. Oranges are removed by the 2 shaker drums revolving horizontally in opposite direction as the spokes free wheeled through the tree canopy. Up to 95 per cent oranges are removed (in small trees) when the machine operated at 1.2 to 2.4 km/hour on one side of the canopy and the harvesting capacity was observed upto 25 tonnes/hour. The fruits fall on a conveyor belt provided below the whorls and trash is removed before filling these in tractor trailer.

2.5 Fruit Harvesters

2.5.1 Banana

The banana harvesting machine is a type of lifting crane (Figure 16). It lifts the bunch from the plant and harvest the bunch from the plant by rotary circular blade. The bunch of fruit is then hung on the rope conveyor for carrying to the packaging hall. Generally a blue colour polythene sleeve is put around the banana bunch for uniform ripening. The polythene sleeve is removed in the packaging hall and bunches are cut into hands and packaged into CFB boxes.

Figure 16: Banana Harvesting Machine

2.5.2 Coconut

The palm climber (Figure 17) designed and developed separately by M J Joseph alias Appachan, P K Joseph and Joy Verghese consists of two metal loops of 10 mm MS rod having sub-loops, rubber belt, wire ropes, connecting clamp, MS plates etc. One loop is intended for the right leg and the other for the left leg. These are called right leg loop and left leg loop respectively, the left loop (main loop) which is slightly bigger than the right loop. The top of the main loop is bent forward to form handle and just below this part, two metal plates are attached with holes with a long rubber belt. On the rubber belt a wire rope having rings on each end is fixed. The bottom most part is a plate and a clamp is provided above it. A long holed plate is fixed on the main loop, which is used as parking brake.

Ropes attached to the climber are used to fasten it to the palm. The ropes passes around the tree and through hooks provided in the climber. A pedal is provided for resting the foot. The right side peddle is raised a little upward so that the grip of the right hand part of the climber becomes loose facilitating upward move-ments by using the right hand and the right leg. Once the right part is moved up the weight of the body rests on the right hand part of the climber. The whole process is repeated for the left hand side. In this way one can easily climb coconut or arecanut palms. The equipment costs about Rs 7,000. The labourer does not feel unsafe when he climbs above 10 meter.

The coconut climber designed by TNAU Coimbtore (Figure 18) has two rectangular iron frames. Each frame is fitted with two rubber bands, which hold the trunk of the tree when weight is put on this. The worker climbs up the tree trunk by

Figure 17: Coconut Climber (Verghese)

Figure 18: Coconut Climber (R-Tech)

transferring his weight on lower frame (by legs) and upper frame (by hip) and lifting the frames alternately.

2.5.3 Grape

Figure 19: Grape Harvester

Self propelled grape harvesters (Figure 19) have become very popular for harvesting wine grapes in California, USA and New Zealand. These harvesters harvest grapes raised on trellis. The grape harvester moves over the row or trellis of grapes. The single trellis picking heads consist of two banks of nylon fingers. The fingers can be straight plastic or spring loaded metal rods or curved rods. The diameter and material of the fingers maximizes the amount of fruit harvested. The vines are shaken by the fingers in the harvesting head. The grapes fall on the collection conveyor and these are then conveyed to the tractor towed grape bin moving in the next row. The trash and leaves are blown by blower. About one to two hectare of grapes is harvested in one hour.

2.5.4 Pineapple

Generally pineapples are harvested manually by a knife and filled into bag or baskets. As the fruit plant has spiny leaves and fruits have crown with spiny leaves, the harvesting labour has to be very skilled. The protective gears like gloves, spectacles and clothing must be worn. The mechanical harvesting machine (Figure 20) is a collecting device for the fruits. Fruits are harvested manually and placed on conveyor belt. The crown and root portion is cut manually before placing the fruit on conveyor belt. The damaged and small fruits are discarded in the field. The conveyor belt is pulled in the field by the transport truck. The conveyor belts carry the fruits to field bins. The pineapple fruits are filled in the field bins by placing the crowns downward to use these as cushion. The harvested pineapples are then taken to the packaging house for grading and packaging. Lights are also provided on the conveyor belt frame for working at night.

Figure 20: Harvesting Pineapples Manually and Putting on Conveyor

3.0 Conclusion

Efforts have been made in India to develop handtools to assist the labour for efficient harvesting of fruits. The man positioning platform has been developed for harvesting of tall trees at Central Institute of Agricultural Engineering, where as tractor mounted lift developed at Indian Institute of Horticultural Research, Bangalore will facilitate mechanical harvesting. These devices are also useful for pruning and spraying of fruit trees. The high capacity self propelled machines developed overseas are required for large orchards. However, we need to import some machines and work on their scaled down versions. When we systematically and scientifically expand the areas under fruit crops even the bigger machines on custom hire basis can become affordable even under Indian conditions.

4.0 Important References

American Society of Agricultural Engineers, 2003. Outstanding Innovatios. *http// www.asae.org/resources on 5.8.2003*

Bucello Vince, 2014. Nelson Citrus and Olive Harvester, Video. *http://www.youtube.com/watch? on 2.5.2014.*

Central Agricultural Engineering Research Institute, 2011. *Development of platform type fruits harvesting system for mango, sapota and citrus*, Annual report 2010-11.

Dole Food Company, Inc, 2013. Harvesting Pineapples. *http://www.dole.com/https://www. youtube. com/watch?v=w FYEeFpvik8, on 3.4.2013.*

Etani Sdn. Bhd.Malasia, 2012. Motorized Oil Palm Cutter, *http://www.etani.com.my/on 2.5.2012*

Harinath, V.N., 2010. Multi-Tree Climbing Devices, A Boon For Toddy-Tappers, The Hindu. *http://www.thehindu.com/on 11.10.2010*

Horticulture Australia Ltd., 2014. Mechanical Citrus Harvester; improvement modifications for commercialisation purposes. *http://www.horticulture.com.au/reports/search_final_reports_ result.asp on 2.5.2014*

Korvan Industries Inc., Oxbo International Corporation, 2010. Oxbo 3220 Citrus Harvester. h*ttp://www.oxbocorp.com/Products/Citrus.aspx http://www.korvan.com, as on 5.10.2010.*

Littau Harvester, 2010. Over-The-Row Harvester. *http://www.littauharvester.com, on 2.5.2010.*

Mandhar, S.C. *et al.* 2003. Mechanization of Fruit Harvesting, **In** *Proceedings of the International Seminar on Downsizing Technlogies for Rural Development* Vol. I,7-9 October, Regional Research Laboratory, Bhubaneswar, India, pp.144-149.

R- Tech Agri Equipments, 2012. Multi Tree Climber. *http://coconuttreeclimber.blogspot.in/p/what-we-offer.html on 21.5.2012*

OESCO, Inc. 2011. Baldwin Wooden Tripod Apple Ladders, *http://www.oescoinc.com/baldwin-wooden-tripod-apple-ladders.html on 14.2.2011*

OMC - Orchard Machinery Corporation, 2011. Shock Wave Tree Shakers, *http://www.shakermaker.com on 11.11.2011*

Parameswarakumar M., and Gupta, C.P. 1991. Design Parameters for Vibratory Mango Harvesting System Transactions of the ASABE. 34(1) 0014-0020

STIHL Incorporated, 2010. Olive Shaker, www.stihlusa.com, on 20.2.2010.

Senthil Kumaran G., 2011. Tractor Trailer Mounted Hydraulic Platform for Harvesting Fruits, *Indian Institute of Horticultural Research,* Bangalore. Personal Communication on 2.7.2011

Verghese Joy, 2011. Kerala Coconut Pluckers, Outlook, *http://www.outlookindia.com/article.aspx?278543, on 17.10.2011*

Zhejiang Zhongli Tools Manufacture Co., Ltd. 2010. Electrical shaker harvester, *http://www.made-in-china.com/showroom/grace 955/companyinfo/Zhejiang-Zhongli-Tools-Manufacture-Co-Ltd-.html on 26.10.2010*

Chapter 5

Pre and Postharvest Management

B.V.C. Mahajan

Professor, Punjab Horticultural Postharvest Technology Centre, Punjab Agricultural University, Ludhiana – 141 004, Punjab, India
E-mail: bvc_mahajan@rediffmail.com

Alemwati Pongener

Scientist, ICAR-National Research Centre on Litchi, Mushahari, P O Ramna, Muzaffarpur – 842 002, Bihar, India
E-mail: alemwati@gmail.com

1.0 Introduction

Plant parts or organs such as fruits, vegetables and flowers derive water, carbohydrates, and nutrients required for growth, development and metabolism, from their parent plant. As long as they are attached to the plant they continue to receive these metabolites. A critical time in the life of these organs is the time of harvest. Once the produce is harvested,it loses its source of nutrients and metabolites, and there is no possibility for further improvement in such components. But the produce, continues to respire and maintain metabolic activities for which it has to rely entirely on its internal reserves of metabolites. The rate at which the harvested produce exhausts its internal reserves, transpiration losses, respiration rate, and various internal and external factors contribute to decide the quality and shelf life of produce.

Quality at the time of harvest is dictated by a multitude of preharvest factors. These factors span from genetic make-up of the crop through cultural and environmental influences to various preharvest treatments. In addition, after harvesting, the produce can be subjected to several postharvest treatments and practices that interplay with the physiological and biochemical processes within the produce to beneficially prolong storage life and maintain quality in fresh form. Extending the postharvest life of horticultural produce requires the knowledge of all pre and postharvest factors that can lead to loss of quality or generation of unmarketable material.

2.0 Preharvest Factors Affecting Postharvest Life and Quality

Several preharvest factors *e.g.*, the rate at which produce deteriorates progress through its normal developmental pattern of maturation, ripening and senescence, affect postharvest life and quality. It is, therefore, very important to understand these preharvest factors. Environmental factors such as temperature, relative humidity, water availability, light, cultural practices and pest management techniques determine the inherent quality of the produce. However, the ultimate quality is the final manifestation of inter-relation between the commodity and its environment.

2.1 Cultivars

Growers have the choice of selecting preferred cultivars based on the eating needs and desires of the consumers. Within each grouping there is a range of genotypic variation in composition, quality, and potential postharvest life. Cultivars vary among themselves for shape, size, productivity, colour, dry matter, taste, ripening time, nutritional quality, shelf life, susceptibility or resistance to pests and pathogens or postharvest handling damages *etc*. Plant breeders have been successful in selecting carrot, sweet potato, and tomato cultivars with comparably high carotenoids levels and vitamin A content, onion and tomato cultivars with longer shelf-life, sweet corn cultivars that maintain their sweetness longer after harvest; cantaloupe and watermelon cultivars with higher sugar content and firmer flesh, and pineapple cultivars with higher contents of ascorbic acid, carotenoids, and sugars. An important limitation of most markets is that only certain cultivars or varieties of a commodity are traded and others are unacceptable. These cultivars receive considerably more attention during postharvest stage while the unacceptable cultivars do not, even though they might constitute a large part of the total production. Growers can, thus, make use of the knowledge of genetic make-up of cultivars with proven superior postharvest qualities for shelf life and quality, and increase market share by replacing the existing genotypes with those of the desirable varieties some such varieties are listed in Table 1.

2.2 Temperature and Light

Temperature and light play important role in improving the growth, yield and postharvest quality of fruits, vegetable and flowers. The temperatures at which most physiological processes go on normally in plants range from approximately 0 °C to 40 °C. For most horticultural crops, higher the temperature during the growing period earlier is the time of harvest. Based on this principle, the heat units or degree-days concept has been formulated. Warm days and cool nights are prerequisites for full development of colour at ripe stage in fruits. Tomato grown at high temperature has low TSS content than those grown in lower temperature due to higher respiration rate at higher temperature. Fruit colouration is also influenced by temperature; cool

Table 1: Different Cultivars of Horticultural Crops with Specific Postharvest Traits

Crop	*Scientific Name*	*Cultivar*	*Attribute (s)*
Apple	*Malus × domestica*	Ambri	Long storage life
		Granny Smith	Firmer than most cultivars
Chilli	*Capsicum annuum*	Punjab Lal	Rich in capsaisin, resistant to TMV, CMV and leaf-curl virus
Fig	*Ficus carica*	Conardia, Excel, and Deanna	Suitable for drying and canning
Grape	*Vitis vinifera*	Pusa Navrang	Tenturier variety suitable for wine and juice making
		Thompson Seedless	Suitable for raisin making
Litchi	*Litchi chinensis*	Swarna Roopa	Resistant to fruit cracking
		Rose Scented	Rose flavour
Mango	*Mangifera indica*	Pusa Surya	Red peel, good shelf life
		Ratna, Arka Puneet	Free from spongy tissue
Strawberry	*Fragaria × ananassa*	Olympus, Hood, Shuksan	High flavour and bright red colour, suitable for ice-cream making

nights with hot dry days favour best colour development. Synthesis of lycopene and colour formation in tomatoes is inhibited at temperatures beyond 30 °C, while, in grapes, there is proanthocyanin accumulation and expression of flavonoid pathway genes in response to diurnal variation in temperature. Curd formation in cauliflower and head formation in cabbage does not take place at temperatures below optimum, while in potato, tuberization does not happen if difference in night and day temperature is not wide.

The intensity and quality of light are both required for optimal plant productivity, quality and harvest index. Availability of light during growth period influences the postharvest eating quality. From nutritional point of view, vitamin-C concentration has direct relation with light intensity in many fruits and vegetables, particularly leafy greens. Tomatoes grown in shade conditions have been found to record higher sugars and dry matter as compared to those grown in open conditions. Another effect of light on fruit quality deals with the red pigmentation of skin, through the influence of light on anthocyanin biosynthesis. Light is a key regulatory factor in reddening of skin in apple and mango.

2.3 Water

Water is indispensable for the growth and development of plants. The availability of water, be it excessive, optimum, or scarce, directly reflects on the yield and quality of plant produce. Severe water stress results in increased sunburn of fruits, irregular ripening of pears, tough and leathery texture in peaches, and incomplete kernel development in nuts. Moderate water stress reduces fruit size and increases soluble solids content, acidity, and ascorbic acid content. On the other hand, excess water supply to plants results in cracking of fruits (*e.g.* cherries, plums, and tomatoes), excessive turgidity leading to increased susceptibility to physical damage (such as oil spotting on citrus fruits), reduced firmness, delayed maturity, and reduced soluble solids content; brittle foliage in leafy vegetables and softer fruits in strawberry with increased tendency to decay. Fruit cracking in litchi and pomegranate is a direct result of dry conditions followed by rain or irrigation. Deficit irrigation or water stress has been reported to reduce yield but increase the soluble solids and postharvest quality in horticultural crops such as tomato, strawberry, ber, kiwifruit and mango. A basic understanding of tree-fruit water balance is important in considering the relationship of plant moisture to fruit quality and shelf life.

2.4 Mineral Nutrition

Nutritional status of fruits, vegetables and flowers is one of the important factors that determine quality of produce at harvest. Optimum plant performance depends on a balanced and timely availability of mineral nutrients that may be limiting in many soils. There can be deleterious effects on storage life and quality of horticultural produce due to deficiencies, excesses, or imbalance of various nutrients during the growing period of the crop. Physiological disorders, as a result of excessive or deficient plant nutrient availability, reduce the quality and marketability of fresh horticultural produce.

High nitrogen content is often associated with reduced postharvest-life due to increased susceptibility to mechanical damage, physiological disorders, and decay in many horticultural crops. Increasing the nitrogen and/or phosphorus supply to citrus trees results in somewhat reduced acidity and ascorbic acid content, while increased potassium fertilization results in increased acidity and ascorbic acid content. High nitrogen also increases the susceptibility of apple flesh to core browning during storage, reduces the shelf life of onion, increases the incidence of soft rot in kiwi fruit and adversely affect the quality of mango fruit by anthracnose disease. High nitrogen application or deficiency of potassium results in inferior fruit quality in grapes, while nitrogen deficiency leads to small fruits with poor flavour. In vegetables, high nitrogen levels delay maturity and increase the incidence of several disorders such as internal browning in tomato, hollow stem of broccoli, increased weight loss during storage of sweet potatoes and soft rots in tomatoes. In general, preharvest treatments with higher dosages of nitrogen are reported to reduce the quality and storage life of the horticultural produce.

Phosphorus is involved in energy transfer, maintenance of cell membranes, and brings about a direct effect on yield and tree health. It affects fruit quality by determining fruit size, firmness, colour and storage potential. The role of phosphorus in storage potential of fruits is attributed to its importance in fruit lipid composition. Excessive phosphorus application can lead to lower soluble solids as well as delayed colour development in citrus. Phosphorus deficiency can lead to delayed maturity, reduction in fruit quality, and decreased disease resistance. Phosphorus application has been found to minimize weight loss, sprouting and rotting of onions.

Among the many plant mineral nutrients, K stands out as a cation having the strongest influence on quality attributes that determine fruit marketability, consumer preference, and the concentration of critically important human health-associated phytonutrients. Fairly high K levels in fruit are often reported to increase red colour and to raise fruit acid level, which can improve fruit taste. However, both high and low levels of potassium have been associated with abnormal metabolism. High levels of potassium have been associated with development of bitter pit in apple. On the other hand, low potassium inhibits lycopene biosynthesis in tomato resulting in delayed development of a full red colour. Application of potassium to citrus trees can affect the shape of the fruits and increase their acidity content.

Calcium is one of the cellular regulators of fruit ripening and deterioration; when calcium is low, these processes proceed more rapidly and fruits are likely to ripen earlier, have greater preharvest drop, more water-core at harvest and develop more scald, brown-core, internal breakdown, internal browning, bitter pit, and rots during and after storage. Of these, only internal breakdown and bitter pit may be the direct symptoms of low calcium. Calcium is often said to be related to fruit softening low calcium levels producing softer fruit, whereas, higher calcium increasing firmness.

Several maladies manifest as a result of excess or deficient macro and micro nutrients. Some important physiological disorders in horticultural crops related to inappropriate nutrition are listed in Table 2.

Table 2: Physiological Disorders in Horticultural Crops Associated with Imbalanced Mineral Nutrition

Element	*Crop*	*Disorder (s)*
Calcium	Apple	Bitter pit, lenticel blotch, cracking, internal breakdown, water core
	Avocado	End spot
	Beans	Hypocotyl necrosis
	Brussel's sprout	Internal browning
	Chinese cabbage	Internal tipburn
	Carrot	Cavity spot, cracking
	Celery	Blackheart
	Cherry	Cracking
	Lettuce	Tipburn
	Lemon	Red blotch
	Mango	Soft nose
	Parsnip	Cavity spot
	Pear	Cork spot
	Pepper	Blossom-end rot
	Potato	Sprout failure, tipburn
	Tomato	Blossom-end rot, blackseed, cracking
	Water melon	Blossom-end rot
Boron	Apple	Internal Cork
	Apricot	Cracking
	Pear	Internal cork, water core

Source: Snowdon, 2010.

While a number of mineral nutrients are essential for plant growth, not all of them have been related to postharvest fruit quality. Market demands high quality fruit. Mineral levels in the fruit as it grows and after it is harvested are highly significant factors in producing and maintaining fruit quality. Many physiological disorders in horticultural crops are attributed to imbalanced mineral nutrition. Both excessive and deficit application of nutrients may have adverse effects on postharvest life and quality of horticultural produce. Maximum benefits can be achieved through integrated nutrient management (INM) as envisaged in the Good Agricultural Practices (GAPs).

2.5 Plant Growth Regulators

The 'classical' phytohormones, identified during the first half of the twentieth century, are auxin, abscisic acid, cytokinin, gibberellin and ethylene. More recently, several additional compounds have been recognized as hormones, including brassinosteroids, jasmonic acid, salicylic acid and nitric oxide.

Plant growth regulators (PGRs) are organic substances, other than nutrients, which in small quantities promote, inhibit, or otherwise modify various plant physiological processes. Every physiological process in the plant can be said to be the result of a delicate balance between such PGRs. Presently, several commercial applications and recommendations are being made to achieve desired effects in plant science, more specifically in crop sciences to achieve better yields. Commercial formulations of PGRs are used in horticultural crops; and PGRs have become so indispensable that any commercial venture in horticulture would not be possible without the use of plant growth substances. There are numerous examples of growth regulators being used commercially to influence some quality attributes of a product. Some compounds such as Naphthalene Acetic Acid (NAA) are used for thinning blossoms in apples. Others such as gibbrellic acid are used to increase fruit size in grape and to maintain rind integrity in citrus. Ethephon is used as a preharvest spray to induce ripening in some crops such as tomato. In order to achieve desired results, it is very important that the correct PGR should be applied in the required concentration at the right time. The commercially followed pre and postharvest applications of PGRs to improve shelf life and quality of horticultural produce are summarized in Table 3.

Plants are used to alter the growth behavior of plants. While vital usefulness of foliar applied growth regulators to fruit trees is generally accepted, however, their use growth regulators does not directly influence fruit composition but may indirectly affect it due to delayed or accelerated fruit maturity.

2.6 Cultural Practices

Soil type, rootstock, mulching, irrigation and fertilization influence the water and nutrient supply to the plant, which can in turn affect the nutritional quality and shelf life of the harvested plant part. Good Horticultural Practices (GHP) ensure good yield, and ultimately the postharvest life and quality of produce. Weeds compete with the main crop for soil moisture, minerals and nutrients, and also act as alternate hosts for diseases and pests. Efficient weed management is an important cultural practice that ensures production of quality horticultural produce. In addition, proper

Table 3: Use of Plant Growth Regulators (PGR) for Enhanced Shelf-life and Postharvest Quality

PGR	*Commodity*	*Concentration*	*Desired Effect*
2,4-D	Citrus, Mango	5-25 ppm	Control of fruit drop
	Limes, Lemon	50-100 ppm	Extending shelf life
	Mandarin	12 ppm	Reduces granulation
Cytokinin	Lime and Lemon	10 and 25 ppm	Extend shelf life
Ethephon	Bael	1000 ppm	Ripening can be advanced by 45 days
	Ber	750 ppm	Induces early maturity
	Date palm	1000 ppm	Advances maturity
	Grapes	100 ppm	Reduce uneven ripening
	Peach cv. July Elberta	300 ppm	Fruit thinning
	Persimon cv. Hachiya	500 ppm	Removal of astringency
	Pineapple	25ppm	Induction of uniform flowering
	Walnut	2000 ppm	Promotes dehiscence of hulls
Ethrel (CEPA)	Apple	1000 ppm	Colour enhancement and enhancement of ripening
Ethrel	Mandarin	50 ppm	Degreening
Ethylene	Avocado	10 ppm	Accelerate ripening
	Banana, Mango, Papaya	100 ppm	Uniform ripening
GA	Grapes	10 ppm	Cluster elongation
	Grapes	40 ppm	Berry thinning
	Grapes	60 ppm	Increasing berry size
	Jamun	60 ppm	Control of flower and fruit drop
	Litchi	40ppm	Reduce fruit cracking
Maleic Hydrazide (MH)	Garlic, Onion	3000 ppm	Sprouting inhibition
NAA	Apple	10-20 ppm	Fruit thinning
	Apple	10 ppm	Control of preharvest fruit drop
	Apricot	25-50 ppm	Fruit thinning
	Grapes	50 ppm	Minimize postharvest berry drop
	Pineapple	10-20 ppm	Induction of uniform flowering
	Pineapple	200-300 ppm	Increases fruit size by 15-20 per cent
Paclobutrazol	Loquat	500 ppm	Increasing fruit size
Promalin	Apple	30-60 ppm	Improving oblong shape of Delicious apples
Putrescine	Peach	5-20 mM	Reduced ethylene production, delayed softening, and retained titratable acidity
Salicylic acid	Sweet cherry	2 mM	Induced disease resistance and reduced disease incidence

Source: Chadha, K.L., *Handbook of Horticulture*, ICAR, New Delhi, 2006.

crop hygiene should be followed to reduce disease outbreak and carryover of postharvest pathogen inoculum. High crop loads can reduce the dry matter in some fruits such as kiwi fruit and peach and may affect the taste and flavour in many horticultural crops. Fruit thinning is done to increase fruit size, which, however, decreases the overall yield. Traditional horticultural technologies, using size-controlling rootstocks and adaptive pruning and thinning techniques, are good examples of enabling technologies to optimize productivity (crop load) and quality of fruit. Berry crops are known to have better postharvest fruit quality when allowed to grow under open canopy with maximum air circulation. Efficient training and pruning practices are recommended to permit sunlight penetration into the center and lower canopy areas. Girdling is a popular practice followed in grape to strike a balance between vegetative and fruit growth, and to improve bunch shape and berry size. Girdling is also done to improve fruit size in peaches and nectarines. Soil-borne fungi, bacteria and nematodes can build up to damaging levels with repeated cropping of a single vegetable crop. To overcome this, crop rotation is an effective management strategy to prevent inoculum build-up in and minimize postharvest losses by reducing decay inoculum right from the field. Effective preharvest disease control greatly influences disease incidence and severity during postharvest handling of fruits and vegetables. Cultural practices have direct effect on the final quality of the produce. Cultural practices such as pruning and thinning determine the crop load and fruit size, which can in turn influence the nutritional composition of fruit.

3.0 Postharvest Treatments for Enhanced Storage Life and Quality

3.1 Thermal Treatments

3.1.1 Hot Water Treatment

Thermal treatment has been used for more than a century to free plant materials from pathogens. Postharvest heat treatments are used for insect disinfestations, disease and decay control of fruits and vegetables. These modify commodity responses to other stresses and maintain fruit quality during storage. In mango, hot water treatment is recommended at 50-52 °C for 5 minutes to reduce the fungal infection during ripening or storage. This treatment helps in attainment of uniform ripening within 5-7 days. Hot water dips have generally been utilized for fungal pathogen control, since fungal spores and latent infections are either on the surface or in the first few cell layers under the peel of the fruit or vegetables. Many fruits and vegetables tolerate hot water temperatures of 50 to 60 °C for up to 10 min, but even shorter exposures at these temperatures can control many postharvest plant pathogens.

An innovation in hot water treatment has been the development of a hot water spray machine. This machine is designed to be a part of a packaging line, whereby the commodity is moved by means of brush rollers through a pressurized spray of hot water. By varying the speed of the brushes and the number of nozzles spraying the water, the commodity can be exposed to high temperatures for 10 to 50 sec. Heat treatment also induces cold tolerance in a number of produce including citrus, pepper and tomato. Broccoli senesces vary rapidly and florets turn yellow after harvest as a result of chlorophyll degradation. This yellowing can be delayed by treating harvested broccoli with hot water (45 °C) for 10 min. However, if the temperature and time is not monitored properly at the time of hot water treatment, it may lead to internal injury to fruit such as flesh darkening in avocado, citrus, litchi and nectarine, and poor colour development, abnormal softening, lack of starch breakdown and development of internal cavities in mango and papaya.

3.1.2 Vapour Heat Treatment (VHT)

Vapor heat, as a method of heating fruit with warm air saturated with water vapour at temperatures between 40 and 50 °C to kill insect eggs and larvae, is used as a quarantine treatment before fresh market shipment. Vapour heat treatment (VHT) proves very effective in controlling infection of fruit flies in fruits after harvest. The boxes are stacked in a room, is heated and humidified by injection of steam. The temperature and exposure time are adjusted to kill all stages of insects (egg, larva, pupa and adult), but fruit should not be damaged. VHT, as a quarantine measure, is mandatory for export of mangoes. VHT can also reduce the incidence of anthracnose during postharvest storage of mango. To achieve complete control of anthracnose, it is necessary to apply a combination of VHT (95 per cent RH and above) such that fruit core temperature of 46.5 °C is maintained for 10 minutes.

3.2 Curing

Curing is performed immediately after harvest on root, tuber and bulb crops, in which the produce is exposed to relatively high temperature and relative humidity. In root and tuber crops, curing causes wound healing with the development and suberization of new epidermal tissues, which act as an effective barrier against infection and water loss. In bulb crops, curing is the process of drying of neck tissues and of the outer leaves to form dry scales. If tubers are placed immediately after harvest into cold storage, wound healing is delayed. Onions and garlic can be cured in the field. These crops can be undercut in the field and left there to dry for 5-10 days. The crops can be cured either in field or after packing into large fiber or net sacks. The produce must be checked daily until outer skin and neck tissues have properly dried. The optimum conditions for curing root, tuber and bulb crops are given in Table 4.

Table 4: Optimum Conditions for Curing Root, Tuber and Bulb Crops

Crop	*Temperature (°C)*	*Relative Humidity (per cent)*	*Duration (days)*
Cassava	30-40	90-95	2-5
Onion and Garlic	27-35	60-75	1-2
Potato	15-20	85-90	5-10
Sweet potato	30-32	85-90	4-7
Yam	32-40	90-95	1-4

3.3 Irradiation

Ionizing radiation can be applied to fresh fruits and vegetables to control micro-organisms, insects and parasites, and inhibit/or prevent cell reproduction and some chemical changes. Food irradiation is a physical means of food processing that involves exposing the prepackaged or bulk foodstuffs to gamma rays, X-rays or electrons. Foodstuffs are generally irradiated with gamma radiation from a radioisotope source, or with electrons or X-rays generated using an electron accelerator. Radiation doses are measured in Grays (Gy). This technology is regarded as a mature and safe procedure, although its approval for use is country and product specific. Internationally, food irradiation has been considered a safe and effective technology by the World Health Organization (WHO), the Food and Agriculture Organization (FAO) and the International Atomic Energy Agency in Vienna. The Codex Alimentarius standard provides that 'the maximum absorbed dose delivered to a food shall not exceed 10 kGy except when necessary to achieve a legitimate technological purpose'. A combination of hot water treatment (55 °C for 5 minutes) followed by 30 Gy irradiation has been found to be the best treatment for shelf-life extension and quality maintenance of mango. After this treatment the fruit exhibits a storage life of 38 days (at 15 °C) with no irradiation injury. Irradiation can also be used to control postharvest diseases of other fresh fruits and vegetables.

In Hawaii irradiation at a minimum of 250 Gy was recommended for the commercial control of fruit fly in papaya, preferably combined with hot water treatment. A dose of 5-10 krad is recommended for inhibition of sprouting in onion and potato. The role of irradiation to reduce postharvest losses, control insect pests and microorganisms, prolong shelf life and improve quality have been tested and found beneficial in a number of horticultural produce, *viz.*, apple, cabbage, carrot, grapefruit, lettuce, mango, papaya, potato, strawberries, tomato, *etc.*

3.4 Biological Control

Increasing public and scientific concerns about presence of synthetic chemicals in food supply and in the environment have provoked the need to look for safer, effective, and more eco-friendly alternatives for postharvest disease control in horticultural crops. Biological control agents or microbial antagonists have been pronounced in control of various postharvest rot pathogens. The yeast *Candida guillermondi* was shown to be antagonistic towards *Penicillium* spp. and could be successfully incorporated into commercial waxes. *Bacillus* spp. on their own or combined with a fungicide could be used to control the postharvest disease of mango, avocado and litchi. *Bacillus subtilis* and *B. licheniform* isolated from mango leaves and applied to fruit as a warm water dip controlled anthracnose levels. Microbial antagonists exert their biocontrol activity through antibiosis, direct parasitism, competition for nutrients and space, or through induction of natural defense mechanisms, such as phytoalexins. Some established reports on postharvest biological control as given in Table 5.

Table 5: Biocontrol Agents Used for Postharvest Decay Control in Horticultural Crops

Commodity	*Disease*	*Biological Control*
Apple	Blue mould	*Pseudomonas syringae, P. cepacia, Cryptococcus* sp., *Pichia guilliermondii*
	Gray mould	*Pichia guilliermondii, Cryptococcus laurentii, C. flavus*
Apricot	Brown rot	*Bacillus subtilis, Pantoea agglomerans*
Citrus	Green mould	*Pichia guilliermondii, Bacillus subtilis,*
	Blue mould	*Pichia guilliermondii, C. laurentii*
	Sour rot	*Bacillus subtilis, Pichia guilliermondii, Trichoderma* sp.
Grape	Gray mould	*Trichoderma harzianum, Pichia guilliermondii*
	Rhizopus rot	*Candida guilliermondii*
Peach/Nectarine	Brown rot	*Bacillus subtilis*
Pear	Blue mould	*Pseudomonas cepacia*
	Gray mould	*Pseudomonas cepacia, P. gladioli*
	Mucor rot	*Cryptococcus laurentii, C. flavus*
Plum	Brown rot	*Bacillus subtilis, Rhodotorula* sp.
Potato	Soft rot	*Pseudomonas putida*
Strawberry	Gray mould	*Trichoderma* sp.

3.5 Waxing

Preservation of fruits and vegetables is a major concern for the fresh produce industry. Extending shelf life translates to the extension of transport distances and, therefore, markets. Edible coatings are a simple technology by which fruits and vegetables can be physically protected and have their respiration and, in some cases, ripening regulated as with passive modified atmospheric packaging (MAP). This is because, like plastic packaging, coatings create a modified atmosphere within the fruit or vegetable. Use of coatings has been practiced for decades in the United State on apples and oranges as well as on other crops.

Waxing is, therefore, a common postharvest practice in those fruits and vegetables which shrivel rapidly and lose consumer appeal during storage and marketing. Food grade waxes are used to replace some of the natural waxes removed during harvesting and sorting operations and can help reduce water loss during handling and marketing. It also helps in sealing tiny injuries and scratches on surface of fruits and vegetables. It improves cosmetic appearance by providing shine and prolongs the storage life of fruits and vegetables.

Materials used for edible coatings include lipids, polysaccharides, proteins, resins or combinations of these materials or composite coatings. Most fruits and vegetables have a waxy cuticle that resists moisture loss, but often this is partially removed or disturbed during washing. This necessitates a coating or packaging treatment to prevent water loss and subsequent shriveling and shrinking of fresh produce. Any 'Generally Recognized As Safe' (GRAS) material that is approved for use in coatings without restriction are considered by the U.S. Food and Drug Administration (FDA) to be "edible".

Edible coatings have been found beneficial on a number of fruit and vegetable crops such as apple, capsicum, citrus, cucumber, plum, tomato, *etc*, to reduce shriveling and improve textural qualities. Shelf life of kinnow and pear fruits can be extended with citrashine coating (Figure 1). Edible coatings have enabled fruits like citrus and apple to be packed and displayed in supermarkets without packaging.

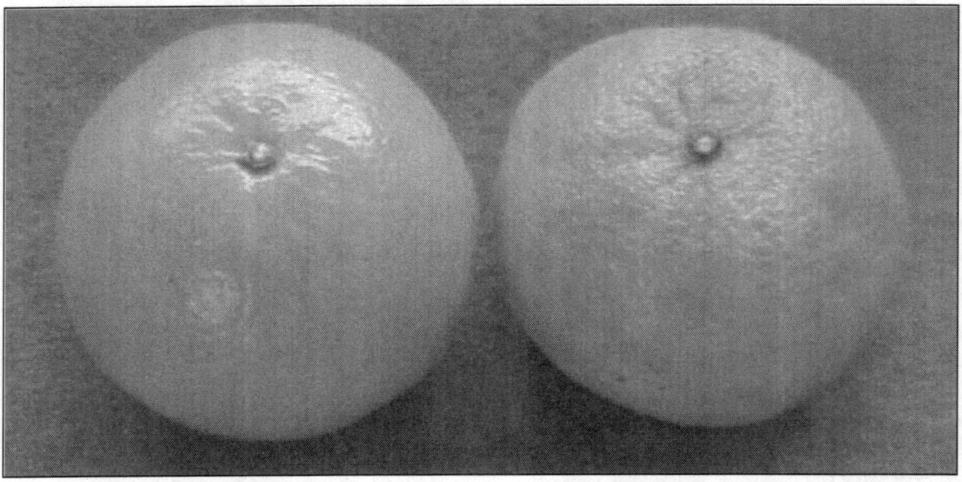

Wax Coating **Without Coating**

Figure 1: Effect of Wax Coating on Appearance of Kinnow Fruits

3.6 Chemical Treatments

3.6.1 Chlorine Solution

Clean water containing an appropriate concentration of sanitizers is required to minimize the potential transmission of pathogens from water to produce, from infected to healthy produce within a single lot, and from one lot of produce to another, over time. Chlorine treatment (100-150 ppm available chlorine) can be used in wash water to help control/reduce microbial inoculum build-up during packing operations. The pH of wash solution should be maintained between 6.5 and 7.5 for best results. Now-a-days, Chlorine dioxide (ClO_2) has been increasingly adopted as an alternative to chlorine for sanitization in the fresh and fresh-cut produce industry. ClO_2 used as a disinfectant has several advantages over chlorine, including larger oxidant capacity, lower reactivity with organic matter, and high effectiveness at low concentrations. ClO_2 is effective as an antimicrobial agent against such bacteria as *Salmonella* and *Escherichia coli* over a broad pH range (3 to 8).

3.6.2 Calcium Application

It has been established that the disorders of storage organs of fruits and vegetables appear closely related to low calcium content in tissues. Postharvest application of calcium plays an important role in enhancing the storage and marketable life of fruits by maintaining their firmness and quality. Calcium application delays aging or ripening, reduces postharvest decay, controls the development of many physiological disorders and increases the calcium content, thus improving their nutritional value. Vacuum infiltration of mangoes with 2-8 per cent $CaCl_2$ solution results in delayed softening of fruit during storage at 20 °C by 8-12 days. Calcium develops resistance to fungal attack by stabilizing or strengthening cells walls, thereby, making them more resistant to harmful enzymes produced by fungi. Calcium dips have been used as firming agents to extend postharvest shelf life of a wide range of fruits and vegetables *viz.* apples, strawberries, sliced cantaloupe, sliced pears, sliced strawberries, blueberries, peaches, tomatoes, oranges, and pineapples.

3.6.3 1-Methylcyclopropene

1-Methylcyclopropene (1-MCP) is a powerful inhibitor of ethylene action that has been added to the list of options for extending the shelf life and quality of fruits. 1-MCP is a cyclic olefin, non-toxic, odourless and stable at room temperature. This compound has been manufactured in powder form releasing 1-MCP when diluted with water. It has been approved by Environmental Protection Agency (EPA), and the commercial application of 1-MCP was undertaken by Agro fresh, Inc., a subsidiary of Rohm and Hass. Presently, this chemical is being traded under the name of Smart fresh™ and several countries are reaping the benefits of this compound. The magic gas, l-MCP, has been used as a tool to enhance the shelf life of perishables, as it irreversibly binds to the ethylene receptors, thereby bringing about a conformational change in the receptors, and prevents ethylene effects in plant tissues for extended period. It has been patented in 1996, and, since then, numerous scientific reports have been documented establishing the role 1-MCP plays in delaying the process of ripening and senescence in fresh horticultural produce (Table 6).

3.6.4 Diphenylamine (DPA)

Diphenylamine is a colourless, aromatic amine possessing antioxidant properties. Postharvest drenching of fruits with DPA provides good control against scald, a serious physiological disorder of apple and pear during storage manifested as brown discolouration that makes the fruit unmarketable and limits its storage potential. In addition, postharvest treatment of pear with DPA prevents normal rates of respiration and ethylene production during ripening and the fruits remain hard and green during storage. DPA reduces the chilling-induced pitting of green bell pepper. DPA is commercially used in USA and Europe on apple and pear fruit for improving their storage life and maintaining quality.

3.6.5 Fumigation

The fumigation with sulfur dioxide (SO_2) is successfully used for controlling postharvest diseases of grape. This is achieved by placing the boxes of fruit in a gas tight room and introducing the gas to an appropriate concentration from a cylinder. Special sodium metabisulphite pads are available which can be packed into individual boxes of a fruit to give a slow release of SO_2. The primary function of treatment is to control *Botrytis cinerea*. The storage life of grape packed along with sulfur dioxide generating pads can be extended upto 50 days at 0-2 °C and 90-95 per cent RH.

Table 6: Effects of 1-Methylcyclopropene on Postharvest Storage Behaviour and Quality of Horticultural Produce

Crop	*1-MCP*	*Altered Postharvest Physiology and Biochemistry*
		a) Fruits Crops
Apple	1 μL L^{-1}	Reduced softening, delayed ripening, inhibited superficial scald and decay
	2 μL L^{-1}	Reduced softening, and reduced superficial scald
Apricot	0.5 μL L^{-1}	Decreased respiration rate and tissue softening
Banana	0.01-1.0 μL L^{-1}	Delayed peel colour change and fruit softening, extended shelf life
Ber	1 μL L^{-1}	Delayed senescence, maintained quality, reduced decay
Cherry	1 μL L^{-1}	Maintained firmness, enhanced shelf life
Grape	0.5-1.0 μL L^{-1}	Increased shelf life, and improved quality
Guava	0.6 μL L^{-1}	Improved shelf life, ameliorated chilling injury, and reduced decay
Kiwifruit	0.5 μL L^{-1}	Suppressed ethylene production and fruit softening
Lime	1 μL L^{-1}	Maintenance of green colour of fruit
Mandarin	5 μL L^{-1}	Inhibited chlorophyll degradation and senescence
Mango	2 mg kg^{-1}	Delayed ripening, increased antioxidant enzymes
Peach	10 μL L^{-1}	Inhibited degradation of starch, prolonged shelf life
Pear	0.3-0.6 μL L^{-1}	Lower ethylene production and delayed softening
Pineapple	0.1 μL L^{-1}	Effectively controlled internal browning or black heart
Plum	0.5 μL L^{-1}	Slowed fruit ripening, reduced fruit softening
		b) Vegetable Crops
Bell Pepper	1 μL L^{-1}	Delayed senescence through decreased chlorophyll breakdown
Brinjal	1 μL L^{-1}	Delays senescence, maintains quality and reduces browning
Broccoli	0.6 μL L^{-1}	Reduced loss of green colour and chlorophyll pigments
Carrot	1 μL L^{-1}	Decreased respiration rate and loss of sucrose
Coriander	0.5 g m^{-3}	Increased shelf life of leaves, retarded chlorophyll degradation
Muskmelon	1 μL L^{-1}	Delayed mesocarp softening and reduced watersoaking
Okra	5 μL L^{-1}	Reduced development of chilling injury and extended storage life
Onion	1 μL L^{-1}	Sprout growth was reduced during storage
Parsley	10 μL L^{-1}	Retard senescence of leaves
Spinach	1 μL L^{-1}	Extended postharvest life, maintained higher chlorophyll
Tomato	5-20 μL L^{-1}	Delayed ripening and increased shelf life
Watermelon	5 μL L^{-1}	Protected against ethylene induced water-soaking
		c) Flower Crops
Carnation	1-20 nL L^{-1}	Inhibited wilting, ethylene effects, and extended vase life four times
Freesia	4 nL L^{-1}	Extended vase life up to 9 days
Lilium	500 nL L^{-1}	Superior quality and inhibition of ethylene response
Phalaenopsis	500 nL L^{-1}	Reduced ethylene induced floral bud drop
Tulip bulbs	200 nL L^{-1}	Inhibited ethylene induced splitting

Source: Watkins, 2008; Schotsmans *et al.,* 2009.

Sulphur dioxide treatment in litchi is commercially practiced by exporters to control pericarp browning. Sulphur dioxide is a well-known inhibitor of polyphenol oxidase and it combines with anthocyanins to form a stable complex. Fumigation with 1.2 per cent SO_2 for 10 minutes was shown to be effective in reducing skin discolouration in fresh litchis, especially if it is combined with a 2 minute dip in 1N HCl directly afterwards. Paper pads or wraps impregnated with biphenyl fungicides are commonly applied to citrus fruits. The chemical vaporizes slowly protecting the fruit from fungal infection.

Despite the good efficacy of SO_2 treatment against decay pathogens, there has been growing concerns against the use of SO_2 in food due to the health issues related to sulfite residues. Concerted efforts are therefore needed to find safer alternatives to SO_2 fumigation for overcoming decay during postharvest storage of horticultural produce.

3.6.6 Sprout Suppressants

In potatoes, a range of chemicals have been shown to suppress sprouting. One such chemical, CIPC (3-chlor-iso-propylphenyl carbamate) is commercially used as a sprout suppressant, which is vaporized and applied as a thermal fog into the stored potato. A periodic fogging of CIPC @ 40 ml/tonne at an interval of 60 days was found optimum for storage of potato at 10±1 °C for 6 months (Figure 2). The potatoes stored at high temperature with periodic CIPC fogging recorded low reducing sugars and negligible sprouting. Sprout suppression in onion bulbs is achieved by preharvest application of maleic hydrazide (MH). This is applied to leaves of the crop some two weeks before harvesting.

CIPC Treated Potato **Untreated Potatoes**

Figure 2: Effect of Postharvest Application of CIPC on Sprouting of Potato

3.6.7 Gibberellic Acid

Gibberellins (GAs) possess the capacity to stimulate growth in many plants. The benefit of the activity of GAs has brought about major advances in the field of physiology and applied plant science, and the use of GAs constitutes one of the most important tools in modern horticulture industry. The role of GAs in delaying the process of senescence in plants and plant tissues is well established. Postharvest application of gibberellic acid (GA_3) has been popularly used as a tool to delay the process of ripening and senescence in various fruits, vegetables and flowers. The anti-senescence effect of GAs is manifested through their effect on delay of colour changes, alleviation of chilling injury, reduction in microbial decay, and increase in storage life of fresh produce. The beneficial use of GA_3 as a postharvest application in various horticultural produce is presented in Table 7.

Table 7: Postharvest Application of GA_3 for Enhanced Shelf Life and Quality

Crop	*Concentration*	*Altered Postharvest Physiology and Biochemistry*
		a) Fruit Crops
Banana	1.5 g kg^{-1}	Delay of fruit ripening
Cherry	10 mg L-1	Higher fruit firmness and lower fungal infection
Grapefruit	100 ppm	Maintained peel quality, reduced decay, and increased storage life
Guava	100 ppm	Delayed onset of senescence and maintained fruit quality
Mango	200 mg L-1	Retarded ripening of fruits
Papaya	100 ppm	Delayed ripening and improved shelf life
Peach	100 mg L-1	Maintained fruit firmness and delayed ripening
Pear	2000 μg mL-1	Improved the efficacy of biocontrol against blue mold rot
	50 ppm	Improved palatability and sensory quality of fruits
Persimmon	200 μg mL-1	Induced resistance to black spot caused by Alternaria alternate
Sapota	150 ppm	Maintained fruit quality and improved shelf life
		b) Vegetable Crops
Celery	20 μg mL-1	Increased resistance to pathogens; inhibited decay
Tomato	0.1 per cent	Extended shelf life up to 18 days
		c) Ornamental Crops
Alstroemeria	100 ppm	Inhibited chlorophyll breakdown and improved vase life
Gerbera	50 mg L-1	Enhanced postharvest quality and vase life of cut flowers
Rose	1 mM	Suppressed the development of Botrytis blight

3.6.8 Jasmonates

'Jasmonates' are a class of endogenous plant growth regulators that have unique and potentially useful properties that affect plant growth and development in response to environmental stresses. Jasmonate is a broad term covering numerous linolenic acid-derived compounds. Postharvest application of MeJA as a gas can activate defences and can positively modify postharvest physiology in a number of crops. Some research on of MeJA on horticultural crops has focussed on preharvest and postharvest treatments to protect against microbial development on harvested tissue. Postharvest exogenous application of MeJA has also been reported to suppress fruit decay, improve quality, and enhance storage life of a number of horticultural crops such as grapefruit, peach, raspberry, and strawberry. Application of MeJA positively reduces the rate of chilling injury in avocado, grapefruit, guava, pepper, pomegranates *etc.* through enhanced antioxidant potential and reduced ion leakage. Applying 200-600 µM MeJA to cut roses (*Rosa hybrida* L.), through the pulsing solution, or as a spray, reduced the incidence and severity of *Botrytis* and enhanced flower colour persistence.

3.6.9 Ethylene Gas

Ethylene, a gas at physiological temperature, is a natural hormone involved in fruit ripening. The autocatalytic production of endogenous ethylene is considered to immediately trigger off the ripening process in fruits like banana, mango, papaya etc. These fruits are commercially ripened by traders with calcium carbide popularly known as *Masala*. This practice is banned under U/S 44-AA of Prevention of Food Adulteration (PFA) Act, 1954 with PFA rules, 1955, due to health reasons. The ripening of fruits with ethylene gas is a safe and internationally acceptable technique. the fruits are exposed to low level of ethylene gas (100 ppm) in an air-tight ripening chamber for 24 hours to induce the fruits to ripen. The most important thing in this technique is temperature and relative humidity control inside the ripening chamber, which should range between 16–24 °C and 90-95 per cent RH, depending upon the fruit type. This technique of fruit ripening is approved by Food Safety and Standard Authority of India (FSSAI), Ministry of Health and Family Welfare. It guarantees uniform ripening with development of good flavour and colour.

4.0 Storage

The goals of postharvest research and extension are to maintain quality and safety and minimize losses of horticultural crops and their products between production and consumption. Providing optimal ranges of temperature and relative humidity (RH) is the most important tool for maintaining quality and safety of intact and fresh-cut fruits and vegetables, herbs, and ornamental crops. Thus, increased precision in temperature and RH management is indispensable for quality maintenance from harvest to final consumption. There is no substitute to maintaining the cool chain throughout the postharvest handling system for maintaining quality and safety of horticultural perishables. All other postharvest technology procedures are supplements to proper temperature management. The optimum storage conditions for horticultural produce are summarized in Table 8.

Every fresh commodity has an optimum temperature at which it must be stored to achieve maximum shelf life. The quicker the temperature is reduced to the optimum storage temperature, the longer its storage life. The process of removal of field heat from the produce is called precooling. Depending on the perishability, nature of the produce, and cost factor, cooling can be achieved through room cooling, forced-air cooling, hydro cooling, vacuum cooling, package icing *etc.* (Figure 3).

5.0 Conclusion

Despite modern interventions and the global introduction of good agricultural practices, postharvest losses remain unacceptably high in both developed and

Table 8: Recommended Temperature, RH and Approximate Storage Life of some Important Fruits, Vegetables and Flowers

Commodity	*Temperature (°C)*	*Relative Humidity (per cent)*	*Approx. Storage Life*
		a) Fruit Crops	
Apple	–1.0-4.0	90-95	1-12 months
Apricot	–0.5-0	90-95	1-3 weeks
Banana	13.0-14.0	85-90	1-4 weeks
Ber	7.5	90-95	2 weeks
Cherries	–1.0 to -0.5	90-95	2-3 weeks
Coconut	0-1.5	80-85	1-2 months
Fig	–0.5-0	85-90	7-10 days
Grapefruit	10-15	85-90	6-8 weeks
Grape	–1.0 to -0.5	90-95	1-6 months
Guava	5-10	90	2-3 weeks
Kiwifruit	–0.5-0	90-95	3-5 months
Lemon	10-13	85-90	1-6 months
Lime	9-10	85-90	6-8 weeks
Litchi	1.5	90-95	3-5 weeks
Mandarin, Kinnow	4-5	90-95	2 months
Mango	13	85-90	2-3 weeks
Papaya	7	85-90	1-3 weeks
Peach	–0.5-0	90-95	2-4 weeks
Pears	–1.5 to -0.5	90-95	2-7 months
Pineapple	7-13	85-90	2-4 weeks
Plum	–0.5-0	90-95	2-5 weeks
Pomegranate	5	90-95	2-3 months
		b) Vegetable Crops	
Asparagus	0-2	95-100	2-3 weeks
Beans, green	4-7	95	7-10 days
Beet	0	98-100	10-14 days
Broccoli	0	95-100	10-14 days
Brussels sprout	0	95-100	3-5 weeks
Cabbage	0	98-100	3-6 weeks
Carrot	0	98-100	4-6 weeks
Cassava	0-5	85-90	1-2 months
Cauliflower	0	95-98	3-4 weeks
Celery	0	98-100	2-3 months
Cucumber	10-13	95	10-14 days
Eggplant	10-11	90-95	2-3 week
Garlic	0	65-70	6-7 months

Contd...

Table 8–*Contd...*

Commodity	*Temperature (°C)*	*Relative Humidity (per cent)*	*Approx. Storage Life*
Ginger	13	65	6 months
leafy Greens	0	95-100	10-14 days
Lettuce	0	98-100	2-3 weeks
Cantaloupe	0-2	95	5-14 days
Watermelon	10-15	90	2-3 weeks
Mushroom	0	95	3-4 days
Okra	7-10	90-95	7-10 days
Onion (green)	0	95-100	3-4 weeks
Onion (dry)	0	65-70	1-8 months
Peas (green)	0	95-98	1-2 weeks
Chilli (green)	0	95-98	40 days
Potato		90-95	5-10 months
Pumpkin	10-13	50-70	2-3 months
Radish	0	95-100	2-4 months
Spinach	0	95-100	10-14 days
Sweet potato	13-16	85-90	4-7 months
Tomato	8-10	90-95	4-7 days
Turnip	0	95	4-5 months
Yams	16	70-80	6-7 months
		c) Flowers	
Alstroemeria	0-1	Sensitive*	1 week
Anthurium	12.5-20.0	Insensitive	1-2 weeks
Gypsophila	0-1	Sensitive	1- 3 weeks
Bird of Paradise	6-7	Insensitive	4 weeks
Carnation	0-1	Sensitive	2-4 weeks
Chrysanthemum	0-1	Insensitive	1-3 weeks
Freesia	0-1	Insensitive	–
Gerbera	0-1	Insensitive	1 week
Gladiolus	0-1	Insensitive	1 week
Iris	0	Insensitive	1 week
Lily (Asiatic)	0-1	Sensitive	Upto 4 weeks
Orchids	0-12.5	Sensitive	–
Rose	0-1	Sensitive	1-2 weeks
Snapdragon	0-1	Sensitive	7-10 days
Tuberose	0	Insensitive	1 week
Tulip	0-1	Insensitive	Upto 7 weeks

Sources for compiling this table include Kader *et al.* USDA Handbook 66, 1992.

* Ethylene Sensitivity.

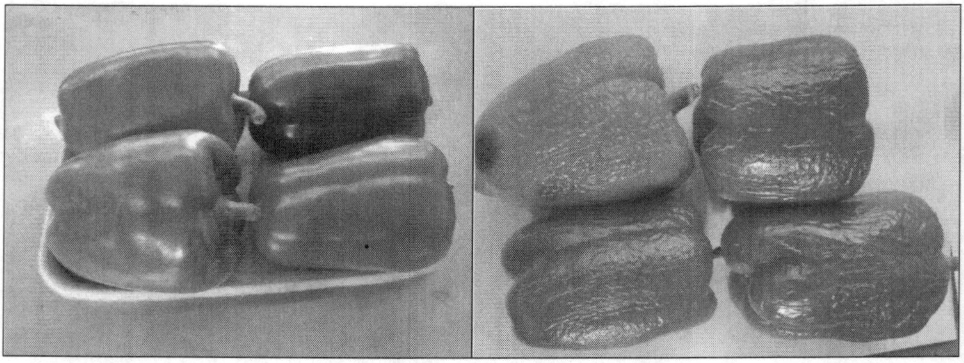

Figure 3: Effect of Shade on Quality of Capsicum

developing countries. New innovative technologies are being developed but effective implementation often remains a hurdle. Much is said about postharvest technology but little has been effectively integrated and implemented, and food security remains a challenge as before. Addressing this requires a paradigm shift in how the problem is viewed and handled.

Assuring optimum postharvest life and quality of horticultural produce is dependent on several pre and postharvest factors. Any produce that is of excellent quality at the time of harvest will not store well if not provided with proper postharvest management practices. Similarly, the best postharvest management technologies would be futile if the produce is of poor quality at the time of harvest. Thus, in order to achieve maximum shelf life and maintain quality of horticultural produce, an integrated approach involving both pre and postharvest factors should be taken into consideration. Best results are obtained when crops are grown under sustainable management techniques, harvested at the right maturity, and provided appropriate combination of postharvest management techniques during subsequent storage and distribution. The storage and marketing of high quality produce requires attention to all aspects of the system, from site and cultivar selection, through all cultural practices during production, to harvest maturity, selection and field handling, grading, packing, temperature management as well as distribution. Produce can be maintained and marketed at high price only if it is of high quality at harvest, and if the postharvest facilities are managed to a high standard.

6.0 Important References

Arvanitoyannis, I.S., Stratakos, A.C., and Tsarouhas, P. 2009. Irradiation applications in vegetables and fruits: A review. *Cri. Rev. Food Sci. Nutri.* **49**: 427-462.

Baldwin, E.A. 2003. Coatings and other supplemental treatments to maintain vegetable quality, p. 413-435. In: J.A. Bartz and J.K. Brecht (eds.). *Postharvest Physiology of Vegetables*, 2nd Edition. Marcel Dekker, Inc., New York, NY.

Chadha, K.L. 2006. Handbook of Horticulture. Indian Council of Agricultural Research, New Delhi.

Cohen, S.D., Tarara, J.M., Gambetta, G.A., Matthews, M.A., and Kennedy, J.A. 2012. Impact of diurnal temperature variation on grape berry development, proanthocyanin accumulation, and the expression of flavonoid pathway genes. *J. Exp. Bot.* **63**(7): 2655-2665.

Cui, N., Du, T., Kang, S., Li, F., Zhang, J., Wang, M., Li, Z. 2008. Regulated deficit irrigation improved fruit quality and water-use efficiency of pear-jujube trees. *Agric. Water Management*, **95**(4): 489-497.

Daane, K.M., Johnson, R.S., Michailides, T.J., Crisosto, C.H., Dlott, J.W., Ramirez, H.T., Yokota, G.T. and Morgan, D.P. 1995. Excess nitrogen raises nectarine susceptibility to disease and insects. *California Agric.* **49**(4): 13-1 7.

El-Samahy, S.K., Youssef, B.M., Aaskar, A.A. and Swailam, H.M.M. 2000. Microbiological and chemical properties of irradiated mango. *J. Food Safety*, **20**: 139–156.

Fallik, E., Grinberg, S., Alkalai, S., Yehkutieli, O., Weisblum, A., Regev, R., Beres, H and Bar Lev, E. 1999. A unique rapid hot water treatment to improve storage quality of sweet pepper. *Postharvest Biol. Technol.* **15:** 25-32.

Gomez-Lopez, V. M., Rajkovic, A., Ragaert, P., Smigic, N. and Devlieghere, F. 2009. Chlorine dioxide for minimally processed produce preservation: a review. *Trends Food Sci. Technol.* **20**: 17–26.

Jin, P., Zheng, Y.H., Cheng, C.M., Gao, H.Y., Chen, W.X., and Chen, H. J. 2006. Effect of methyl jasmonate treatment on fruit decay and quality in peaches during storage at ambient temperature. *Acta Hort.* **712**: 711-716.

Kader, A.A. 1992. *Postharvest Technology of Horticultural Crops*. 2nd Edition. University of California, Division of Agriculture and Natural Resources. Publication, pp. 3311.

Kays, S.J. 1999. Preharvest factors affecting quality. *Postharvest Biol. Technol.* **15**: 233-247.

Kester, J.J. and Fennema, O. 1986. Edible films on coatings: A review. *Food Technol.* **40**(12): 47-59.

Léchaudel, M. and Joas, J. 2007. An overview of preharvest factors influencing mango fruit, quality and postharvest behaviour. *Braz. J. Plant. Physiol*. **19**(4): 287-298.

Lurie, S. 1998. Postharvest heat treatments. *Postharvest Biol. Technol*. **14**: 257-269.

Mahajan, B.V.C., Dhatt, A.S. and Dhillon, W.S. 2004. Effect of prestorage treatments on the quality and storage life of Asian pear. *Indian J. Hort.* **61**: 324-26.

Mahajan, B.V.C., Dhatt, A.S. and Sandhu, K.S. 2005. Effect of postharvest treatments on the storage life of Kinnow. *J. Food Sci. Technol.* **42**: 296-99.

Mahajan, B.V.C., Dhatt, A.S., Sandhu, K.S. and Garg, A. 2008. Effect of CIPC on the storage and processing quality of potato. *J. Food Agri. Environ.* **6**: 34-38.

Pearce, B.D., Grange, R.I., and Hardwick, K. 1993. The growth of young tomato fruit. I. Effects of temperature and irradiance on fruit grown in controlled environments. *J. Hort. Sci.* **68**: 1-11.

Purvis, A.C. 2002. Diphenylamine reduces chilling injury of green bell pepper fruit *Postharvest Biol. Technol.* **25**(1): 41-48.

Rohwer, C.L., and Erwin, J.E. 2008. Horticultural applications of jasmonates: A review. *J. Hort. Sci. Biotechnol*. **83**(3): 283-304.

Sayyari, M., Babalar, M., Kalantari, S., Martinez-Romero, D., Guillen, F., Serrano, M., and Valero, D. 2011. Vapour treatments with methyl salicylate or methyl jasmonate alleviated chilling injury and enhanced antioxidant potential during postharvest storage of pomegranates. *Food Chem.* **124**: 964-970.

Schotsmans, W.C., Prange, R.K., and Binder, B.M. (2009). 1-Methylcyclopropene: mode of action and relevance in postharvest horticulture research. *Hort. Reviews*. **35**: 263-313.

Sisler, E.C., and Blankenship, S.M. 1996. Method of counteracting an ethylene response in plants. U.S. Patent 5518988.

Sisler, E.C., and Serek, M. 1997. Inhibitor of ethylene responses in plants at the receptor level: Recent developments. *Physiol. Planta.* **100**: 577-82.

Snowdon, A.L. 2010. *Postharvest diseases and disorders of fruits and vegetables*. Manson Publishing Ltd. London, UK.

USDA (2014). *The Commercial Storage of Fruits, Vegetables, and Florist and Nursery Stocks*. Agriculture Handbook Number 66 (HB-66). Agricultural Research Service, Beltsville Area/BARC. http://www.ba.ars.usda.gov/hb66//

Watkins, C.B. (2006). The use of 1-methylcyclopropene (1-MCP) on fruits and vegetables. *Biotechnol Advances*, **24**: 389-409.

Watkins, C.B. (2008). Overview of 1-Methylcyclopropene trials and uses for edible horticultural crops. *Hort. Sci*. **43**(1): 86-94.

Chapter 6

Postharvest Diseases

R.D. Rawal

Former Head, Division of Plant Pathology,
ICAR-Indian Institute of Horticultural Research,
Hessaraghatta, Bengaluru – 560 089, Karnataka, India
E-mail: rawal.rd@gmail.com

Jyotsana Sharma

Principal Scientist,
ICAR-National Research Centre on Pomegranate,
NH-9, Solapur-Pune Highway, Kegaon,
Solapur – 413 255, Maharashtra, India
E-mail: jyotisharma128@yahoo.com

1.0 Introduction

Horticultural crop production in India has revolutionized agriculture sector during the last two decades. Postharvest handling and associated losses due to diseases, however, are major challenges in storage, transport and export. Diseases may reduce quantity as well as quality of the harvested produce. The extent of postharvest losses in horticultural crops depend not only on postharvest handling and storage conditions but also on preharvest conditions of the crop in the field. Horticultural practices adopted including cultivation practices, plant protection measures, plant nutrition and environmental conditions prior to harvest contribute significantly to postharvest losses due to diseases. Virulence of the causal organism, and susceptibility of the commodity (the host), the storage atmosphere (temperature, relative humidity and composition of the storage air/atmosphere), stage of produce maturity, postharvest hygiene adopted, extent of damage (cuts/bruises/punctures) caused to the produce during handling and treatment methods, all contribute to the extent of losses due to postharvest diseases. The most important factors in maintaining quality and extending the shelf-life of fruit and vegetables after harvest are temperature and relative humidity. Most of the physical, biochemical, microbiological and physiological reactions contributing to deterioration of produce quality are largely dependent on temperature and humidity. Due to their high moisture content and nutrient composition, fruits and vegetables are very susceptible to attack by different pathogens. Inoculum loads (population) of the pathogen on the host and in the storage atmosphere are other important factors responsible for postharvest losses. Among different postharvest pathogens fungi and bacteria are major etiological agents. Fungi are the main cause of most postharvest diseases barring a few where bacteria are involved. Fungi are highly potent parasitic organisms. A single species of fungus like *Rhizopus stolonifer* is able to infect more than three hundred species of fruits, vegetables and other plants. Viral agents are not the major cause of postharvest diseases; however some preharharvest viral pathogens may develop faster in the storage conditions.

Although it is very difficult to determine the full extent of postharvest losses due to decay of fruits and vegetables, which varies widely with commodity, production area, season, growth conditions, storage conditions and length of storage, it is well known that these losses are significant, particularly in under-developing and tropical countries. In India, postharvest losses have been estimated to be more than thirty per cent. This not only results in negative impact on our economy but also hampers the efforts in providing food and nutritional security to the masses.

Today, the consumers are health conscious and fruit or fruit products are in great demand. Consumers require products that are not only of optimal nutritional quality but are also safe for human consumption, hence, postharvest handling especially treatments to control postharvest diseases assume special significance. It is therefore, of utmost importance to work out effective and economically viable management strategies to reduce postharvest losses.

Postharvest diseases may be categorized as (i) latent/quiescent infections and (ii) those that arise due to infections during or after harvest. In quiescent infections, disease initiates before harvest in the field, but remains dormant for some time in storage. With biochemical changes in the stored produce, the pathogen again becomes active, leading to infection. *Colletotrichum* sp. and *Botrytis cineraria* are two such postharvest pathogens of several commodities. The second type of infection is caused by pathogens like *Penicillium, Rhizopus* and *Erwinia* spp. We summarize here major postharvest diseases of commercial fruits and vegetables and their management.

2.0 Postharvest Diseases of Fruits

A wide range of microorganisms attack fruits during postharvest period. Among different pathogenic microorganisms, fungi are the major organisms deteriorating fruits after harvest. Bacterial pathogens are not common in fruits due to their acidic nature. Postharvest diseases of some important crops are discussed below.

2.1 Apple

2.1.1 Diseases

More than 20 fungal pathogens have been found associated with postharvest losses in apple from Himachal Pradesh only. Losses between 10 to 18 per cent have been reported. The calculated yearly losses due to fungal pathogens ranged from 3.06 to 20.6 thousand tonnes in Himachal Pradesh during 1969-1977. The cumulative loss for the decade was thus estimated at 89.15 thousand tones. The major postharvest pathogen of apple (Figure 1) is *Penicillium expansum* contributing to major losses followed by other pathogens discussed below.

a) Bitter Rot

This disease is incited by *Glomerella cingulata* (Ston) Spaud. and Schrenk. The disease is particularly severe on Golden Delicious, however, Ruspippin and Yellow Newton varieties are also infected occasionally. The disease appears as a faint, light brown discolouration beneath the skin. The discolouration increases in surface diameter and extends in a cone shaped manner. The rough, circular lesion becomes depressed (Figure 1a). Tiny, raised, black dots appear beneath the cuticle. They give rise to acervuli which break the cuticle. In moist weather, sticky, pink masses of spores are seen which may be arranged in a clearly defined rings. Infected fruits of Richard Red and Golden Delicious show decrease in quantities of amino acid, organic acid and carbohydrates. Trisodium phosphate (2 per cent) solution can check the disease in storage.

b) Core Rot

The core rot incited mainly by *Alternaria alternate,* disease is commonly found in apples having open pore at the calyx end. Typical rot symptoms of the disease are nearly round; brown to black lesions, often centered around a skin break or weakened tissue. The spots are firm, dry and shallow. The surface of spots becomes dark brown to black in the advanced stages, the rotten tissue becomes spongy and the affected flesh turns black (Figure 1b).

c) Pink Mould Rot

This rot incited by *Trichothecium roseum* Link. Ex Fr. is prevalent and destructive. The internal infection becomes evident by a pinkish fluffy growth of the fungus. The rot is usually found associated with scab. In the beginning, rot appears as sunken; brown bands about 1/8 inch wide encircling the scab spots. On the rotten area, fungal threads appear and under moist warm conditions pink spore masses are produced. The second or final stage of rot is characterized by chocolate brown, sunken area with an irregular outline. The rotten area remains rather firm and dry. The affected tissue has a bitter taste. In the absence of apple scab, the rotting begins from any point of injury on the fruit surface. The rotten flesh becomes musty. The seed cavity contains a pink spore mass.

d) Soft Rot

This disease is incited by *Penicillium expansum* and is responsible for major postharvest losses in India. In the early stages, the rotten area appear as light in colour and are soft and watery in texture. The rot advances rapidly. In humid atmosphere, grey blue cushions of fruiting bodies of the pathogen appear on the diseased skin. Moreover, the attack of *P. expansum* may lead to the accumulation of patulin - a mycotoxin whose ingestion may cause serious health problems (Figure 1c).

e) Whisker's Rot

This disease is caused by *Rhizopus arrhizus* Fischer. The disease pathogen incites a soft rot. The flesh becomes sour in taste without any undesirable odour (Figure 1d).

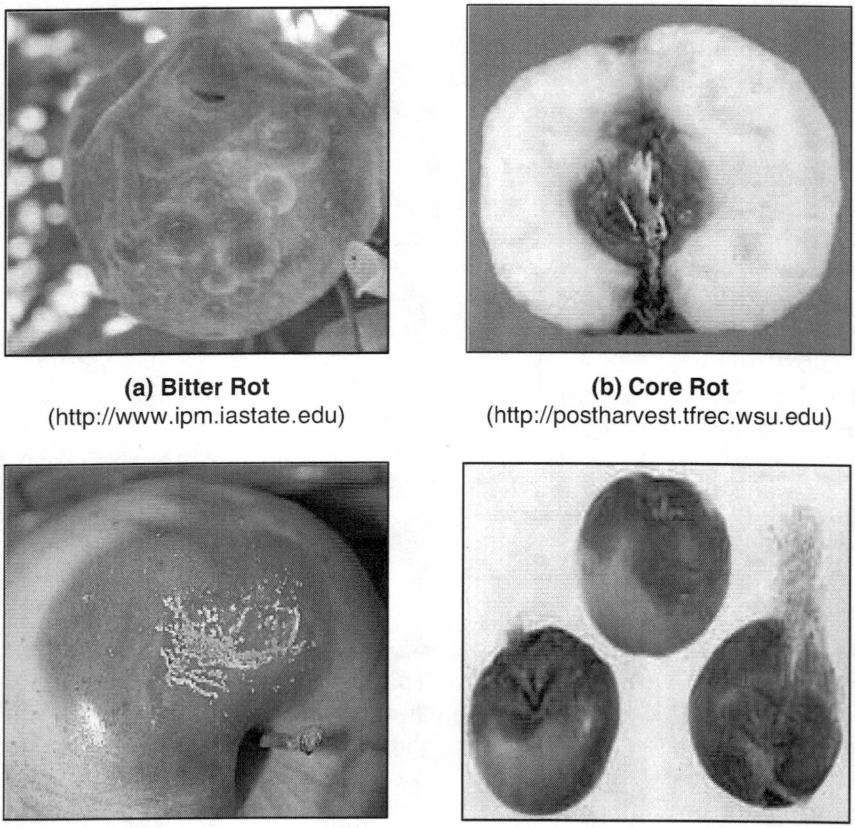

(a) Bitter Rot
(http://www.ipm.iastate.edu)

(b) Core Rot
(http://postharvest.tfrec.wsu.edu)

(c) Soft Rot
(http://botit.botany.wisc.edu)

(d) Whisker Rot
(http://postharvest.tfrec.wsu.edu)

Figure 1: Postharvest Diseases of Apple

2.1.2 Disease Management

Postharvest diseases in apple are due to either contamination or incipient infections that occur during the growing season or due to infection through injuries attributable to harvesting, processing, packing and transporting of the produce. The skin injury should be avoided by observance of stringent precautions in picking, grading and packing. The diseased and injured fruits should be sorted out and discarded. Over ripe fruits should not be stored. Fruits should be placed in cold storage after packing. To control the postharvest diseases the fruit temperature should be kept below 10 °C. Growth and spread of the rots are very slow at low temperature and may be almost entirely inhibited at 0 to 1.6 °C.

Several postharvest treatments are applied to optimize the effect of low temperature *e.g.* sulphurdioxide fumigation by use of sodium metabisulphite. Apples coated with paraffin oil and pungent mustard oil withstand rotting. Fruits wrapped with SOPP impregnated wrappers showed some protection as well. Benomyl at 1000 ppm was the best single treatment for control of rot at about 25 °C. Dipping fruits in 1000 ppm diphenyleamine for 5 minutes or in 500 ppm aureofungin for 20 minutes resulted in the best control of the rots. Pathogenic isolates most resistant to thiabendazole are sensitive to imazalil. Use of 2-aminothiozole (5 per cent) plus thiourea (5 per cent) has been reported for best control of *Rhizopus* rot. Maximum protection is achieved by using fumigation with sodium metabisulphite and wrapping with dichloronitroaniline wraps. Trisodium phosphate 2 per cent solution can check the bitter rot disease in storage. Treatment of fruits with thioacitamid suppressed development of symptoms of core rot up to 35 days. *Pseudomonas syringae* is recorded to be most effective biocontrol against blue mould pathogen *P. expansum*.

2.2 Banana

2.2.1 Diseases

The postharvest pathogens in banana generally gain entry into the skin of unripe fruit through bruises, wounds or injuries, but remain quiescent until the fruits start ripening. The pathogens then multiply rapidly resulting in blackening and softening of tissues. The major postharvest diseases of banana (Figure 2) are caused by fungal genera *Collectotricum, Gloeosporium, Fusarium, Botryodiplodia* and some other minor pathogens.

a) Anthracnose

It is one of the most important and widely distributed disease of maturing bananas caused by *Collectotricum musa*. Symptoms include anthracnose, black spot, stem and crown rot. The infected spots on the surface of the fruits are initially light, moist, but later change in colour to black. Relative humidity of above 85 per cent favours both conidial germination and development of anthracnose on ripe bananas. A high nitrogen dose and total sugars and a low content of ascorbic acid in the peels increases susceptibility to the pathogen (Figure 2a).

b) *Botryodiplodia* Fruit Rot

It is one of the most severe storage pathogens of banana causing heavy losses in Uttar Pradesh, Maharashtra and Andhra Pradesh. The disease is also called as stem end rot and is caused by *Botryadiplodia theobromeae.* The fungus is a wound parasite. Infection originates in or immediately below the decayed perianth or style and spreads uniformly along the fruit causing a progressive brownish black discolouration. The rotting of pulp is comparatively faster than that of the skin (Figure 2b).

c) Cigar End Rot

In India the disease is incited by *Verticillium theobromae.* Fungal pathogens such as *Gloeosporium musarum* and *Verticillium* have also been reported to cause cigar end rot. Infection generally begins from the tip of the finger and slowly spreads backwards causing greyness of the skin. The affected area later gets covered with the white mycelia over which black coloured fructification appear. It becomes corrugated and appears like a burnt cigar tip. Fruits get infected when they are still immature. In advanced stage the pistil end of the fruit shows cracks and the pulp inside the infected zone becomes dry and powdery. Some varieties belonging to Cavendish group of Banana are more susceptible to this pathogen (Figure 2c).

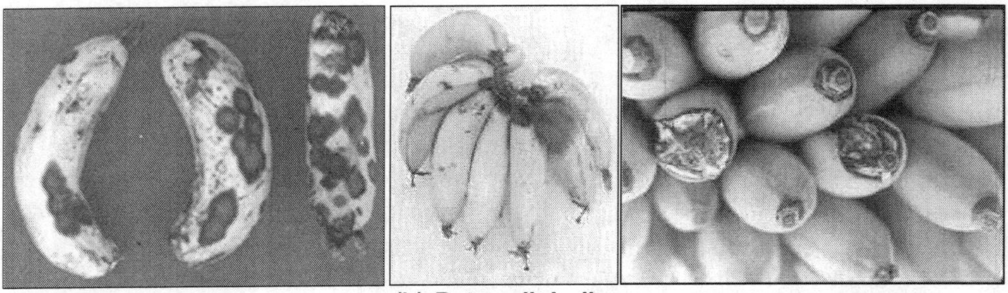

(a) Anthracnose **(b) *Botryodiplodia* Rot** **(c) Cigar End Rot**

Figure 2: Postharvest Diseases of Banana

d) *Fusarium* Fruit Rot

Rotting of banana fruits by *Fusarium moniliformae* was first reported from Allahabad and is now prevalent throughout the country. Olive brown spots appear on infected fruit. The spots extend longitudinally and coalesce. The fruit surface turns clove brown or mummy brown. The fruits become pulpy, ooze out a liquid and emit undesirable odour. Later the rotten tissues get covered with white cottony growth of the fungus.

e) Pink Mould Rot

The pink mould rot is caused by *Trichothecium roseum* and has been reported from Uttar Pradesh, Delhi, Andhra Pradesh and Bihar. The rot is characterized by pink mould growth over the surface of the fruit. The rot remains limited to peel. The fungus is a week parasite and has a wide host range.

2.2.2 Diseases Management

The young bunches should be opened up for more light and air so as not to create conducive environment for the pathogens. Cigar-end rot can be checked to some extent by removing of pistil and perianth by hand.

Most of the fungal pathogens require temperature between 25-30 °C. Storage at low temperatures reduces postharvest decay. As bananas are sensitive to temperatures below 13 °C, it is advisable to store bananas between 13-15 °C to slow down decay.

Moreover, bananas have short storage life, hence, need to be transported and marketed at the earliest.

Preharvest sprays of banana with carbendazim (0.1 per cent) and benomyl (0.1 per cent) reduce latent fungal infection in the field. Prophylactic postharvest dip before storage with benomyl (0.1 per cent) and aureofungin (100ppm) give complete control of postharvest rotting due to fungi. Two preharvest sprays of Burgundy mixture (0.3 and 0.5 per cent) or Bordeaux mixture (0.5 and 1.0 per cent) on 60-75 days old fruits gives best control. Preharvest spray with azoxystrobin 23 per cent SC (0.06-0.1 per cent) or difenoconazole 25 per cent EC (0.015 per cent) 15 days before harvest followed by dipping in hot water at 42 °C has been reported to effectively check fungal diseases like anthracnose. Bio-agent *Burkholderia cepacia* against anthracnose disease and *Trichoderma viride* against pink mould rot have been reported to check postharvest decay.

2.3 Citrus

2.3.1 Diseases

Citrus fruits have a delicate rind which is liable to injuries and wounds in field as well as after harvest, hence requires proper handling at all stages. Most of the pathogens attacking citrus are moulds like *Penicillium, Aspergillus, Botrytis etc.* which are weak parasites and require injury/wound for entry. In addition some other pathogens also lead to serious losses if postharvest handling and management are not proper. Important postharvest diseases of citrus (Figure 3) are described below.

a) *Alternaria* Fruit Rot

Black rot incited by *Alternaria citri* Ell. and Pierce has been noticed in Punjab, West Bengal, Rajasthan and Madhya Pradesh. The rot develops between 15 to 30 °C. On naturally infected fruits, small, discoloured spots appear which turn black. External injury to rind in var. Malta leads to infection and disease development, which increases with the extent of injury. The core of fruit becomes black. The fungus reduces soluble solids and reducing sugars and increases percent acidity in the affected fruits (Figure 3a).

b) Anthracnose

Colletotrichum gloeosporioides infection is known for causing anthracnose in oranges and grape fruits and stem end rot in citrus fruits. Infections usually start from stem end as brown necrotic spots, which gradually spread towards blossom end and cover upper half or the entire fruit in severe cases. This type of rot is observed more in case of Coorg mandarin during main season and to a lesser extent during monsoon crop.

c) *Aspergillus* Rot

In India, 10 to 15 per cent of lime fruits are affected by *Aspergillus niger.* The disease is usually noticed in July – August. The rot begins as small, circular water soaked spots which enlarge and turn brown. The soft rot destroys the whole fruit within 8-10 days. The rotten fruits emit a fermented odour. Contact with the diseased fruits can infect injured as well as healthy fruits. Percentage of rotting is greater at 30-35 °C, particularly when humidity is high (Figure 3b).

d) Brown Rot

This disease is common in mandarin oranges grown in heavy rainfall areas of South India *viz.* Coorg, Wayanad and parts of Palni hills. The disease is caused by *Phytophthora nicotianae.* Apparently healthy fruits carry latent infection and develop brown soft fruit rot during storage (Figure 3c).

e) *Penicillium* Fruit Rots

The *Penicillium* moulds occur in all the citrus growing countries. In India, their occurrence was first reported by Cheema *et al.* (1937). The fruits are affected both in transit and storage. The rind of affected fruit becomes watery on pressing and breaks easily. The entire area gets covered with the green or blue mouldy growth. Finally the entire fruit rots and emits a foul smell. The blue mould rot is incited by *Penicillium italicum* Wchmer and the green mould rot by *Penicillium digitatum* (Pers.:Fr.) Sacc Fruits weakened by long cold storage decay more readily than fresh fruits (Figure 3e).

f) Stem End Rot

Diplodia and Phomopsis stem end rot are the two most important fungal diseases of fruits encountered in transit. The earliest symptom is a softening of the rind around the button. The affected area increases rapidly and becomes brown. The discolouration is usually darker and a tendency to show decay at both ends is a characteristic of *Diploidia* stem end rot. Peak decay occurs 2-3 weeks after harvest (Figures 3f).

2.3.2 Disease Management

Careful handling and avoiding injury to the soft rind help in prolonging storage life. The rot can be minimised by avoidance of injury through sorting and discarding infected fruits and by storing fruits at low temperature in baskets or open containers. Good results can be obtained by pruning and burning the diseased twigs in winter and by spraying with 1 per cent Bordeaux mixture in February, March and September. Out of various chemicals tried, prochloraz and benomyl give the best results in controlling storage diseases. Copper fungicides (1 per cent) and karathane (1 per cent) as preharvest sprays can also check infection and disease development.

A five minute dip in 2 per cent bleaching powder and 10 minute dip in 0.2 per cent captan followed by storage at 2-4 °C have been reported effective against both the moulds. Thiabendazole was found to be markedly ineffective in reducing the decay as its protective quality remained for a short period. Thiabendazole at 0.1 per cent incorporated in waxol-0-12 has been found useful under refrigeration. Boric acid (4 per cent) is effective against green mould. postharvest treatment with 6-8 per cent borax solution or 300-400 ppm mycostatin is recommended for stem end rot. Imazilil WP and Imazilil EC formulations were highly effective in reducing decay and showed reduction index of 74.16 to 96.66 during a storage period of 36 days. Postharvest application of lipid based hormone jasmonates reduced decay caused by *P. digitatum* in "Marsh Seedless" grapefruit.

(a) *Alternaria* Fruit Rot
(http://www.plantprotection.hu)

(b) *Aspergillus* Rot
(http://www.shouragroup.com)

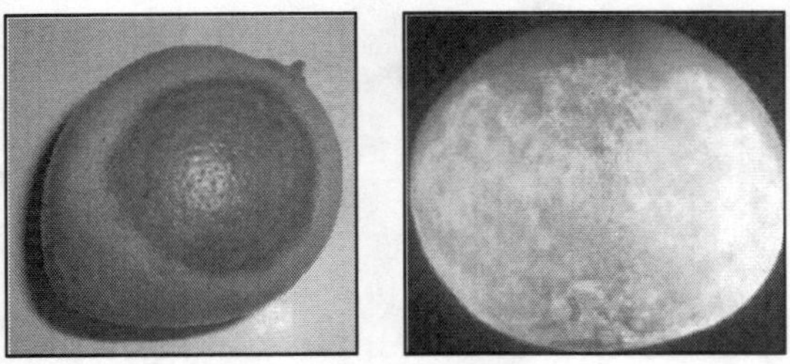

(c) Brown Rot
(http://point09acres.blogspot.in and http://www.shouragroup.com)

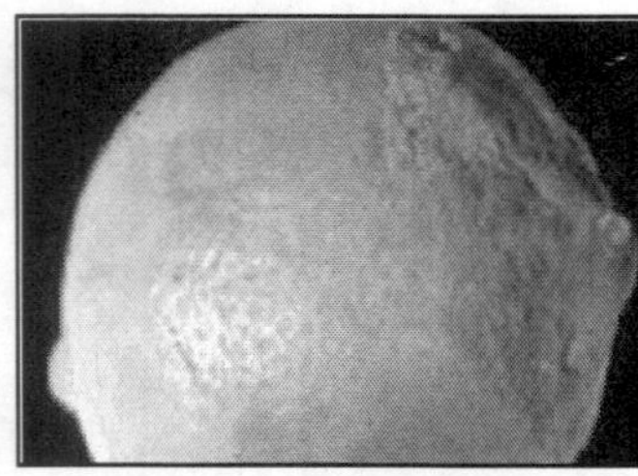

(d) *Colletotrichum* Stem End Rot
(http://www.daff.qld.gov.au)

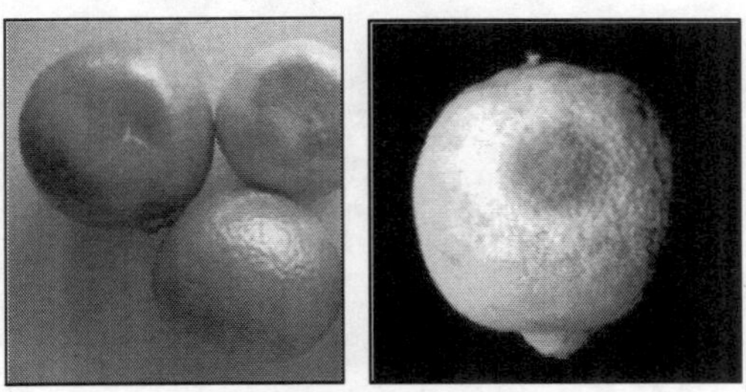

(e) *Penicillium* Fruit Rot
(http://www.pestnet.org and http://plantdiseasehandbook.tamu.edu)

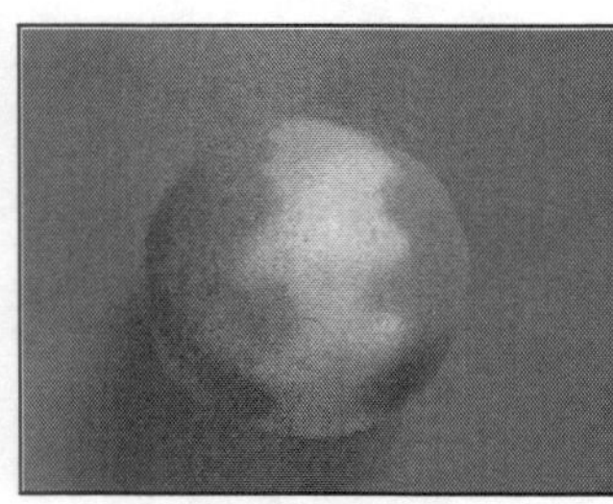

(f) Stem End Rot
(http://edis.ifas.ufl.edu)

Figure 3: Postharvest Diseases of Citrus

Control of decay of citrus fruit caused by *Penicillium spp.* was also reported with bacterial antagonists - *Bacillus subtilis* (Ehrenberg) Cohn, *Burkholderia* (*Pseudomonas*) *cepacia* Palleroni and Holmes, and *Pesudomonas syringe* Van Hall and yeast, *Kloeckera apiculata*(Rees) Janke and *Pichia anomala* (Hansen) Kurtzman. Bacterial suspensions of *Erwinia chrysanthemi* RK-67 and *Bacillus subtilis* RK-6 are reported to reduce rot severity due to *Aspergillus* sp. on lemon cultivars.

2.4 Grape

2.4.1 Diseases

Grapes being succulent are susceptible to attack by several rot causing fungi like *Botrytis, Penicillium, Alternaria, Cladosporium, Aspergillus, Greenaria, etc*. The important postharvest diseases of grape (Figure 4) are described here.

a) Bitter Rot

The disease produces brown, hard, depressed patches on the berries, which subsequently bear numerous black, dot like fructifications. Harvested grapes with small scars or lesions deteriorate in transit and storage showing typical fruit rot symptoms. The disease is incited by *Greenaria uvicola* (Figure 4a).

b) Blue Mould Rot

As the name suggests the disease is incited by blue mould fungus- *Penicillium digitatum* and is characterised by scanty growth of the pathogen, which is at first white but turns bluish green after sporulation. The affected tissues become soft and watery and develop mouldy flavour (Figure 4b).

c) *Cladosporium* Rot

This disease is incited by *Cladosporium herbarum*. The pathogen causes a black, firm, shallow decay, which is usually localised on one side or blossom end. The fruit becomes flat or wrinkled on the affected side. A sparse surface growth of grey-green fungus may be seen on the fruits kept at room temperatures (Figure 4c).

d) Grey Mould Rot

Botrytis cinerea Pers.: Fr. (teleomorph: *Botryotinia fuckeliana* (de Bary) Whetzel) is a plant pathogenic fungus distributed world-wide and causes grey mould disease on both preharvest and postharvest plant tissues. The disease on grapes is incited by *Botrytis cinerea*. In compact bunches, berries on the inside of the attached bunch may split during growth and may be invaded by fungal pathogens resulting in bunch rot. Even a single infected berry can cause 'nest rot' during transit and storage because the pathogen can spread by contact. Mature fruits are more

(a) Bitter Rot

(b) Blue Mould Rot (http://agritech.tnau.ac.in)

(c) *Cladosporium* Rot (www. archive.agric.wa.gov.au)

(d) Grey Mould Rot (http://www.agf.gov.bc.ca)

Figure 4: Postharvest Diseases of Grapes

susceptible to *B. cinerea*. Losses due to this mould on grape are dependent on harvest date and environmental conditions during or prior to harvest. Bunches picked after rains are more prone to decay (Figure 4d).

2.4.2 Disease Management

The rapid growth of pathogens resulting in rotting of grapes can be reduced by preventing field infection through suitable preharvest sprays, by careful handling to reduce bruises and injuries, precooling and refrigeration. Postharvest fumigation is recommended to reduce inoculum load of the pathogen, followed by well planned marketing of stored lots according to their decay potential. Fruits for transit should be precooled to at least 4 °C and then stored at 0.55 – 0 °C. This will delay decaying of berries and reduce postharvest losses.

Sodium bisulphite which releases SO_2 when in contact with air can be used with packing material to control storage diseases. *Vinifera* grapes fumigated with SO_2 in storage, reduce losses due to various *rots*.

Preharvest infection can be controlled by three applications of captan 50WP (0.25 per cent) at monthly interval before rainfall. Benomyl is equally effective as a preharvest spray. Postharvest dip with captan (0.2 per cent) and difolatan (0.2 per cent) reduced berry rot above 80 per cent due to *Cladosporium*, showing respectively 12.84 and 10.11 per cent rotten berries as against 78.80 per cent in untreated control.

2.5 Guava

2.5.1 Diseases

Various organisms associated with storage losses in guava are *Phomopsis psidii, Phoma psidii, Macrophoma allahabadensis, Phytophthora nicotianae var. parasitica, Pestalotiopsis psidii, Colletotrichum gloeosporioides, Cylindrocarpon tonkinense, Rhizopus stolonifier, Curvularia tuberculate, Geotrichum candidum, Botryodiplodia theobromae, Aspergillus niger, Trichothecium roseum*, etc. Some of the important pathogens leading to postharvest rot in guava (Figure 5) are discussed here.

a) *Botryodiplodia* Fruit Rot

This disease is incited by *Botryodiploidia theobromae,* which causes considerable damage during storage and transit. The affected fruits show brownish discolouration mostly at the stem end which generally proceeds downwards in an irregular wavy manner. In advanced stages, numerous small pycnidia are produced over the entire surface of the fruit. The rot is soft and watery.

b) *Colletotrichum* Fruit Rot

This is a serious postharvest disease in north east and other rainy parts of India and results in pin-head sized spots on the unripe fruits which enlarge to form sunken and circular, tan to dark brown to black spots. Infected tissue in unripe fruit becomes hard and corky (Figure 5a).

c) Fruit Canker

The fruit canker is widely prevalent in India and incited by *Pestalotiopsis psidii* and *Colletotrichum gloeosporioides*. Infection generally occurs on green fruits. The canker does not penetrate deep into the flesh, but the market value of severely infected fruit is greatly reduced. Small brown, circular, necrotic brown areas with raised margin appear on fruit which later break open epidermis. Safeda variety is highly resistant to canker (Figure 5b).

d) *Macrophoma* Fruit Rot

Different species of *Macrophoma* are known to cause diseases in mango, banana, *and other fruits.* A severe fruit rot caused by *Macrophoma allahabadensis* was first described from Allahabad markets.The disease occurs in the field as brown blotches on the fruits which leads fruit to rot. Sucrose is completely hydrolyzed after infection and D-glucose and D-fructose are either consumed or reduced in concentration. Marked changes take place in free and bound amino acids, amides and organic acids

e) *Phoma* Fruit Rot

Phoma rot incited by *Phoma psidii* is characterised by brown coloured spots produced over the fruit surface. They ooze out cream coloured pycnidiospores. Sucrose in the inoculated fruits is completely hydrolyzed in six days (Figure 5c).

f) *Phytophthora* Fruit Rot

Fruit rot is incited by *Phytophthora nicotianae* var. *nicotianae* and is known to occur in Karnataka, Tamil Nadu, Andhra Pradesh, Maharashtra and Rajasthan. Moderate temperature, high humidity and rainy weather favour the disease incidence. Infected area becomes greyish brown and appear water soaked with greyish black center.

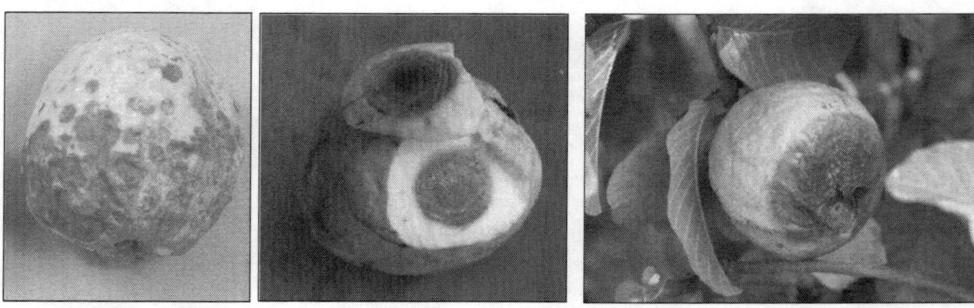

(a) *Colletotrichum* Fruit Rot **(b) Fruit Canker** (http:// www.insectimages.org) **(c) *Phoma* Fruit Rot** (http://agritech.tnau.ac.in)

Figure 5: Postharvest Diseases of Guava

2.5.2 Disease Management

As in other fruit crops preharvest operations and environmental conditions, postharvest handling and cold storage are all the more important in guava for deciding extent of postharvest damage. Cold storage and avoidance of fruit injury are the best means of reducing postharvest decay. Careful handling and storage at 15 °C or below and quick disposal minimise the losses.

Control of most rots can be achieved by 6 sprays at monthly intervals with dithane Z-78 (0.2 per cent) or aureofungin (10 ppm) during fruiting. thiram, copper based fungicides, ziram 27 per cent SL and NC 1584 are good protectants in pre inoculation sprays. Field sprays of chlorothalonil or mancozeb at 15 day intervals give good control of the fungal diseases both in the field and in the storage. Sprays of thiophanate methyl and tricyclazole in the field have been found to control this disease effectively. The rot can be checked by treating the fruits with Thiobendazole (TBZ). Some varieties like 'Safeda'are highly resistant to canker, which may be preferred in areas where canker is a problem.

2.6 Mango

2.6.1 Diseases

In India fungi belonging to around 20 genera are known to attack mango fruits. In addition, some bacteria are also known to decay mango fruits in stores. The important pathogens are *Botryodiplodia theobromae, Colletotrichum gloeosporioides, Rhizopus arrhizus* and *Aspergillus niger*. Some other pathogens of minor importance in mango include - *Alternaria tenuissima* (Alternaria rot), *Sclerotium rolfsii* (soft rot), *Boothiella tetraspora* (dry rot), *Actinodochium jenkinsii* (Black spot), *Pestalotiopsis mangiferae* and *P. glandicola* (Brown rot), *Phomopsis mangiferae* (Fruit rot), *Guignardia mangiferae* (Black rot) and *Pseudomonas mangiferae* (Bacterial rot). The major postharvest rots (Figure 6) are discussed here:

a) Anthracnose

Anthracnose of mango is caused by *Colletotrichum gloeosporioides* and results in loss of produce before and after harvest. The exporters and farmers are equally concerned about this disease. It appears as black or dark brown spots on the fruits. In advanced stages of the disease, the fungus produces asexual fruiting bodies called acervuli with abundant orange to salmon pink masses of conidia on the lesions. The pathogen can grow at temperatures above 15 °C, the optimum being 25 °C (Figure 6a).

b) Black Mould Rot

The rot is caused by *Aspergillus niger*. The affected fruits show brown to pale yellow colour on the damaged part of the fruit with the development of irregular, hazy, greyish spots which coalesce into dark brown or black lesions. The mesocarp of the rotten area gets depressed and softens over which smoky black coloured mouldy growth is conspicuous. Variety 'Totapuri' is highly susceptible to *Aspergillus* rot (Figure 6b).

c) Soft Rot

The pathogen *Rhizopus arrhizus* incites a typical soft watery rot. The disease develops rapidly between 20 and 40 °C. Development of disease is slow and of low intensity at low temperatures (7-8 °C).

d) Stem End Rot

Botryodiplodia theobromae earlier name *Diplodia natalensis* is an important postharvest disease of ripe mango adversely affecting the export of mango. In the initial stages, the epicarp darkens around the base of the pedicel. In the next few hours, the affected area enlarges to form a circular black patch, which under humid conditions extends rapidly and within 2–3 days turns the whole fruit completely black. Profuse black coloured pycnidial sporulation is predominant on the infected part of the fruit. The pulp of the diseased fruit becomes brown and soft. The pathogen incites the soft rot with the aid of pectinolytic and cellulolytic enzymes. The infected fruits lose ascorbic acid rapidly and have lower non-reducing sugar contents in the pulp (Figure 6c).

(a) Anthracnose **(b) Black Mould Rot** **(c) Stem End Rot**

Figure 6: Postharvest Diseases of Mango

2.6.2 Disease Management

Dead twigs and bark of mango trees harbour several types of pathogen. With the onset of rains, the atmosphere of the mango orchard gets contaminated with spores of the pathogenic fungi. These are carried to the store as latent infection on the fruits. Hence, orchard sanitation is of utmost importance. Dead discarded twigs and plant debris should be collected and burnt, instead of dumping in or around the orchard.

Preharvest sprays with chemicals like carbendazim (0.1 per cent) and prochloraz (0.1 per cent) are effective in reducing rot in storage. Since carbendazim is under review and prochloraz is not easily available in India, the preharvest sprays of difenoconazole and tricyclazole are nowadays recommended for controlling the disease. In addition formulations of Milcop, Fycol, copper oxychloride also have been recommended as sprays. Carbendazim (0.1 per cent), benomyl (0.1 per cent), thiophanate methyl (0.1 per cent) and bitertanol (0.1 per cent) spray at 15 day interval, 4 times before harvest, significantly reduces postharvest rotting in mango. However, preharvest interval should be taken into consideration to minimize pesticide residue in the fruits.

Short exposure to temperature of 40 °C kills several fungal pathogens, hence, hot water treatment of harvested mangoes at 50 to 55 °C for 15 minutes is recommended before storage. The treated fruits can be stored for 25 to 47 days at 3.5-10 °C. Dipping mangoes in 6 per cent borax solution at 43 °C for 3 minutes can also reduce rot infection. Good control of the disease is achieved by dipping of fruits in Benomyl (1500 ppm) and thiobendazole (900 ppm). Among the effect of various postharvest treatments after preharvest sprays, prochloraz (0.1 per cent) and benomyl (0.1 per cent) dip and hot water treatment at 52 °C for 10 minutes give effective control of the diseases.

Studies on efficacy of antagonists indicate that four isolates of bacteria, five yeasts and two filamentous fungi are effective *in vitro* against anthracnose (*Colletotrichum gloeosporioides*) in mango. *Brevundimonas diminuta* isolate B-62-13, *Stenotrophomonas maltophilia* L-16-12, a member of Enterobacteriaceae L-19-13, *Candida membranifaciens* F-58-22, and the yeast isolate B-65-23 are effective on naturally infected fruit. The isolates significantly reduce severity of anthracnose on artificially inoculated mango fruit. Biocontrol of black mould rot by *Trichoderma harzianum* has been reported.

2.7 Papaya

2.7.1 Diseases

Papaya is affected by 8 different types of rots,. *Alternaria* fruit rot incited by *Alternaria alternata* and *Botryodiplodia* fruit rot incited by *Botryodiplodia theobromae* are other postharvest diseases of papaya which need more investigation. Over the years these pathogens have become serious and need concerted efforts to work out their management strategies. The important postharvest diseases of papaya are described below in Figure 7.

a) Anthracnose

Colletotrichum gloeosporioides is very common in all the papaya growing countries. The disease occurs as lesions on the fruit initially as brown, superficial discolouration of the skin. The spots soon develop into circular, slightly sunken areas, which may be water-soaked. The lesions may coalesce and white mycelial growth appears on the margin of spots (Figure 7a).

b) *Ascochyta* Rot

Ascochyta caricae Pat. causes both fruit and trunk rot. The disease generally occurs on half grown or mature fruits that show small, circular water soaked spots. The lesion may become sunken and brownish-black. The rot usually spreads outwards irregularly. Rotting becomes most severe in high temperature and high humidity.

c) *Fusarium* Fruit Rot

Fusarium solani causing fruit rot is reported from Bengal and other parts of northern plains. Affected fruits show small, stem end or surface lesions. Profuse, white hyphae are seen associated with the rot.

d) *Marcrophomina* Fruit Rot

A fruit rot of papaya incited by *Marcrophomina phaseolina* is responsible for considerable postharvest loss. The disease is characterised by small, water soaked spots, which appear as circular specks on the fruit surface. The spots rapidly enlarge and become sunken. The pathogen advances deep into the fruit and causes rotting and disintegration. The affected fruits turn brown to black. The attacked tissue hardens and become dotted with sclerotia of the pathogen.

e) *Phomopsis* Fruit Rot

Fruit rot caused by *Phomopsis caricae* pv. *papavae* in papaya develops a water spot, which increase in size. The rotten area depresses and cracks at a later stage. The affected portion turns dark brown to black. Water soaking is seen in the area surrounding the depressed portion. In some fruits, the rotten portion is surrounded by raised tissue, white in colour (Figure 7b).

f) *Rhizopus* Rot

Fruit rot in papaya caused by *Rhizopus stolonifer* occurs in different parts of India. The rot is characterised by the production of irregular water soaked lesions which gradually enlarge and get covered by white and dark brown fungal growth and sporangiophores. Finally the fruit collapses by becoming watery and emits a foul odour. The rot appears when cold stored fruits get exposed to environmental heat during retailing (Figure 7c).

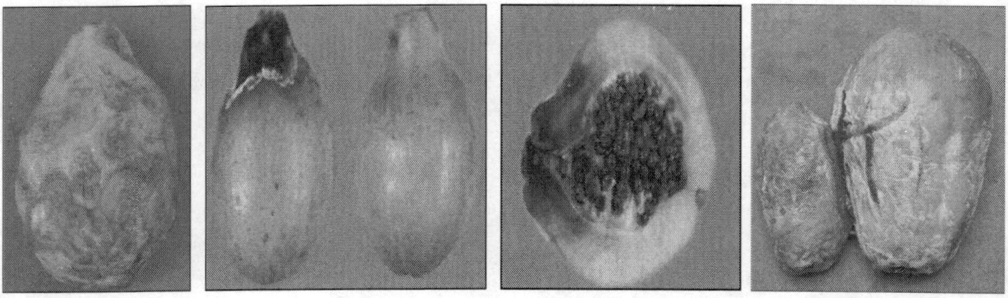

(a) Anthracnose **(b) *Phomopsis* Fruit Rot** (http://agritech.tnau.ac.in) **(c) *Rhizopus* Rot**

Figure 7: Postharvest Diseases of Papaya

2.7.2 Disease Management

Combination of preharvest sprays, postharvest dips and hot water treatments give best control of papaya rots. Hot water double dip treatment (42 °C for 30 minute followed by 20 minute at 49 °C) in combination with preharvest sprays of prochloraz or tricyclazole and storing the fruits at 10 °C or below, have been found to be extremely effective in managing the storage diseases. Spraying the fruits with Bordeaux mixture (2:2:50) at intervals of 21 to 30 day from fruit setting stage, controls several pathogens. Two field sprays either with bavistin (0.1 per cent) at 45 day interval or with daconil (0.2 per cent) at 15 day interval were sufficient to minimize anthracnose after harvest. Preharvest sprays of thiophanate methyl or difenoconazole provide a perfect control of postharvest rots. The combined effect of prochloraz at 0.1 per cent as preharvest spray and prochloraz 0.1 per cent as postharvest dip are also highly recommended in controlling the anthracnose disease of papaya. The incidence of *Ascochyta* rot can also be reduced by fruit dip in 2-aminobutane carbonated solution (1 per cent a.i.) or by fumigation with 2-aminobutane. Borax, ziram, captan, and dithane M-45 are effective in preventing symptom development when applied to the detached fruits. Benomyl and difolatan though reduce rot are advised to be avoided for use in the field before harvest and postharvest use. *Trichoderma viride* and *Pseudomonas fluorescence* have been found to suppress *Marcrophomina* fruit rot.

2.8 Pineapple

2.8.1 Diseases

Pineapple is not affected by many postharvest pathogens, probably because of its tough hard outer rind, which protects from injuries and bruises, thus restricting avenues for entry of rot fungi which are generally weak pathogens. It has only one serious pathogen *Ceretocystis paradoxa* causing stem end rot.

a) Stem End Rot

The stem end rot, core rot or black rot incited by *Ceretocystis paradoxa* (*Thielaviopsis paradoxa*) is a major fungal pathogen which attacks pineapples after harvest and is the chief cause of deterioration of the stored fruits (Figure 8). The infection generally starts at the cut end of the stem in the form of small circular, water soaked spots, which are very soft. With the advancement of disease the spots enlarge and coalesce forming a large black patch extending throughout the fruit. The inside tissue is also invaded by the fungus, which becomes very soft, black, watery and emits foul smell (Figure 8).

Figure 8: Stem End Rot of Pineapple

2.8.2 Disease Management

Chemical control trials conducted have shown that dipping the fruits for 5 minutes after harvest in thiobendazole (1000 ppm) or benomyl (2000 ppm) would ensure good protection against this disease. Dusting of the stem end with a small amount (0.1 g per cent) of benzoic acid with an inert substance such a kaolin, is highly effective in controlling the stem end rot and is also commercially feasible on account of the low cost of the treatment.

2.9 Pomegranate

2.9.1 Diseases

Important postharvest disease of pomegranate are rots caused by *Phomopsis punicae Rhizopus arrhizus, R. stolonifier Aspergillus nidulans, Aspergillus clavalus, Spicaria sp., Aspergillus niger, Syncephalastrum racemosum Botryodiplodia theobromae, Coniella granati* and *Penicillium expansum*. Most of the rot pathogens in pomegranate may gain entry through calyx end and sometimes from stem end or injuries on rind. Important postharvest diseases of pomegranate (Figure 9) are described here.

a) Black Blotch

It is caused by *Botryodiplodia theobromae.* It causes severe fruit rot and spoils the entire fruit including arils. The disease is characterized as black blotch on any part of the fruit which spreads very fast and under favourable conditions covers the entire fruit. The dark coloured pycnidial fruiting bodies are the typical characteristic of the disease and the pathogen. The disease is most serious in the post monsoon season.

b) Black Mould

It is caused by *Aspergillus niger* and is a pathogen of concern which occurs frequently in all seasons and under all storage conditions. The pathogen enters through injuries or calyx end. The pathogen results in brownish soft rots, with a depressed centre in later stages where blackish sporulation of the fungus can be seen. These generally may grow inside without any external symptoms for some time. The rind and underlying pulp become soft, but fruits retain their shape (Figure 9a).

c) Dry Rot

It is caused by *Colletotrichum gleosprioides, Botryodiplodia theobromae* and *Coniella granati*. These are mostly preharvest infections which may develop in storage as hard dry brown rot. *Colletotrichum* rots develop as tan brown dry rot usually from calyx end. Arils also are affected but rot is dry. This rot is very common in pomegranate growing belts in India, while other species are more important in other countries (Figure 9b).

d) Heart Rot

This disease is basically a preharvest problem, but remains unnoticed till fruits are cut open. The pathogen enters the host at the time of flowering and develops heart rot symptoms later during fruit maturity or in postharvest period. The infection by the pathogen - *Alternaria alternate* results in black rot of arils, without any exterior symptom. In advance stages the rind may become slightly off colour or paler red (Figure 9c).

e) Other Rots

Some of the fruit rot organisms such as *Cladosporium oxysporum, Aspergillus flavus, A. niveus* and *A. versicolour, Penicillium chrysogenum,* produce orange-red, dull brown, black slimy, soft watery patches on pomegranate fruit, which often emit fermented smell/ foul odour

f) *Phomopsis* Fruit Rot

Pathogen *Phomopsis punicae* causes fruit rot disease, symptoms of which start from calyx end as dark brown zonate rings and gradually spread over the entire fruit. Pycnidia of the pathogen appear on the affected area. In severe cases the aril also gets rotten and makes the fruit unmarketable. When such fruits with minor infection are packed in a box and transported, adjacent fruits also develop typical rotting within a week.

g) Soft Rot

Soft rot is the major impediment in pomegranate storage and is incited by *Rhizopus arrhizus* and *R. stolonifier*. It starts as small spots which increase in size and coalesce. The disease appears on the rind but the entire fruit also decays internally into a pulpy mass. Under dry conditions, cracking originates from the point of infection.

(a) Black Mould (b) Dry Rot (c) Heart Rot

Figure 9: Postharvest Diseases of Pomegranate

2.9.2 Disease Management

Timely disease control and sanitation (removal of old fruit and dead branches), reduce the incidence of the disease. Infected, fruit drops to the ground by gently shaking the tree at the time of harvest. Care should be taken during harvesting and packing to avoid injury to the fruit skin. Treatment of packing straw with SO_2 producing chemicals can protect the fruit from soft rot. Satisfactory control of soft rot has been achieved by treatment of fruits with linseed, mustard and castor oil. Protecting fruits from physical or insect injury save the fruit from rotting to a great extent. Harvested fruits should be stored at 5-6 ℃ with minimum 80 per cent RH after precooling.

The preharvest sprays during crop season with carbendazim (0.1 per cent), propiconazole 0.1 per cent thiophanate methyl (0.15 per cent), mancozeb (0.2 per cent), difenoconzole 25 EC (0.05 per cent), captan (0.2 per cent), benomyl (0.2 per cent), ziram (0.25 per cent), Bordeaux mixture (0.5-1 per cent), copper hydroxide (0.15-0.2 per cent) and copper oxychloride (0.25 per cent) take care of most of the postharvest diseases. Combination treatments of carbendazim (0.1 per cent) + mancozeb (0.2 per cent) and benomyl (0.1 per cent)+ mancozeb (0.2) control all postharvest pathogens and also improve fruit quality. The formulation containing tricyclazole 18 per cent +mancozeb 62 per cent WP@2.5g/l has been found effective in reducing *Colletotrichum* rot. Fruit dip with dithane Z-78 (0.1 per cent), captan (0.2 per cent), benomyl (0.2 per cent) also reduces postharvest rots. The damage of *Phomopsis* rot can be greatly reduced by 3 sprays at 10-day interval with 0.5 per cent copper hydrate plus albolium oil.

2.10 Strawberry

2.10.1 Diseases

Botrytis and *Colletotrichum* are the major pathogens of harvested strawberry (Figure 10). Other fungi capable of infecting strawberries include species of *Rhizopus, Penicillium, Fusarium* and *Alternaria*.

a) Anthracnose

Another important postharvest disease of strawberry is caused by *Colletotrichum fragerae*. The disease appears as depressed salmon coloured spots on mature and ripe fruits, with abundant pink coloured spore mass oozing out. If one such fruit goes in a box overnight more than fifty percent fruits get infected and they become unmarketable (Figure 10a).

b) Grey Mould

Botrytis (grey mould) is the most important fungus that has been detected in strawberry fruit samples limiting its storage life. Studies show that *B. cinerea* is a necrotrophic pathogen. The fruits disintegrate and grey mould growth is seen on infected fruts. It infects plant tissues through multiple mechanisms, including secretion of cell wall degrading enzymes (*e.g.* cutinase and polygalacturonases) and phytotoxic metabolites (*e.g.* botrydial and oxalic acid) (Figure 10b).

(a) Anthracnose
(http://www.omafra.gov.on.ca)

(b) Grey Mould
(http://www.apsnet.org)

Figure 10: Postharvest Diseases of Strawberry

2.10.2 Disease Management

Though many effective chemicals have been reported to control postharvest pathogens in other crops, these cannot be used in strawberry, as the strawberry fruit is consumed immediately and raw, To overcome this problem many microbial agents have been tried and found to be effective in suppression of *B. cinerea* on numerous crops under both preharvest and postharvest conditions. Commercial products based on *Candida oleophila*, such as Aspire (strain I-182), on *Metschnikowia fructicola*, such as "Shemer",and on *Trichderma harzianum*, such as Trichodex (strain T39), are now available for protection of postharvest fruits from infection by *B. cinerea*. Exogenous putrescine may have an anti-pathogenic function in strawberry fruits enhancing the shelf life of strawberry. The preharvest spray with tricyclazole, difenoconazole and thiophanate methyl keeping in mind the PHI of these chemicals to avoid residues, followed by the treatment with *Trichoderma harzianum* is the most effective to check rot.

3.0. Vegetables

Most of the vegetables are perishable commodities and hence, require careful handling and storage. Postharvest diseases of vegetables are caused by microscopic fungi and bacteria. Vegetables are less acidic than fruit hence, are more vulnerable to bacterial infections than fruits. Postharvest diseases of important vegetable crops are discussed below.

3.1 Beans

3.1.1 Diseases

Many rot fungi including *Sclerotinis sclerotiarum, Botrytis cinerea, Colletotrichum lindemuthianum, Phytophthora.capsici* and *Pythium* are known to cause postharvest diseases in beans (Figure 11). Watery soft rot and grey mould-rot are the most common maladies affecting beans.

a) Anthracnose

Anthracnose caused by *Colletotrichum lindemuthianum* is characterised by deep black spots with reddish brown border. One to many, isolated or coalesced spots may be present on the affected pods Pink spores are found on the centre of the spots (Figure 11a).

b) Cottony Leak

It is caused by *Phytophthora capsici* and infects the pods. Most of the time the fungus is not as fluffy as white mould, moreover there is a reddish border between healthy and infected pod tissue (Figure 11b).

c) Grey Mould

The disease results in soft rot on bruised tissue with *Botrytis cinerea* growth on affected parts. The mould spreads to neighbouring beans and spoils them (Figure 11c).

d) Pod Rot

Pythium pod rot is a problem during wet, warm weather. The pathogen is soil borne and pods near ground are attacked/contaminated first which spread the disease to other pods in field as well as in the stores. Occurrence of this disease is extremely hard to predict, unless the field has a history of *Pythium* problems (Figure 11d).

(a) **Anthracnose** (http:// www.cals.ncsu.edu) (b) **Cottony Leak** (https://extension. udel. edu) (c) **Grey Mould-Rot** (http:// mtvernon.wsu.edu) (d) **Pod Rot** (https:// extension.udel.edu)

Figure 11: Postharvest Diseases of Beans

3.1.2 Disease Management

Refrigeration at 4.4 – 7.2 °C immediately after harvest prevents spread of most of the diseases in store. Field sprays of Ridomil Gold or Copper are effective to check bean diseases.

3.2 Cabbage

3.2.1 Diseases

The major postharvest diseases in cabbage are caused by fungal pathogens *Botrytis cinerea, Sclerotinia, Alternaria* and bacterial pathogens *Erwinia carotovora* and *Xanthomonas* (Figure 12).

a) Bacterial Zonate Spot

The disease is caused by *Pseudomonas cichorii* (Swingle) Stapp. First symptoms of the disease are light brown, 2-13 mm diameter irregular lesions on the cabbage head, later turning darker brown. The spots give target board effect. Infected areas are firm and rubber like. It is an important market disease and is sometimes confused with *Alternaria* rot due to zonate effect. Curd rot due to infection of *Alternaria brassicae* is very common in cabbage, broccoli and cauliflower.

b) Black Rot

It is a serious problem of cabbage and is caused by *Xanthomonas campestris*. It is a vascular disease, which blackens the veins. It shows vascular discolouration at core of stored cabbage. It is primarily a field disease characterized by darkening of vascular elements forming black network of leaves and also leads to losses in store (Figure 12a).

c) Grey Mould

The disease is caused by *Botrytis cinerea* and is a serious disease in prolonged storage under moist conditions. The rot occurs as dark lesion on the outer leaves, and progresses inwards until the whole head is infected. The pathogen appears as a grey fluffy fungal mass, on the outside of the cabbage and as a brown soft rot penetrating deep into the head. Untreated cabbage can lead to losses above 40 per cent and worsens with long-term refrigerated and Controlled Atmosphere stores (Figure 12b).

d) Soft Rot

The disease caused by *Erwinia carotovora* is another serious problem in cabbage and cauliflower. Infection enters in the field through stumps and may remain unnoticed for long time. The infected area inside heads rots under favourable conditions. Rotted parts look cream coloured and later turn brown with foul smell. Injury also helps weak parasites like *Rhizopus stolonifer* to infect cabbage causing soft rot (Figure 12c).

(a) Black Rot (http://vegetablemdonline.ppath.cornell.edu)

(b) Grey Mould (http://www.farminguk.com)

(c) Soft Rot (http://www.agroatlas.ru)

Figure 12: Postharvest Diseases of Cabbage

3.2.2 Disease Management

For control, affected heads or leaves should be discarded during packaging. Use maximum care to reduce bruising of heads. Close trimming of infected leaves before shipment helps reduce removing latent infection.

The best control is storage under high relative humidity and low temperature with good circulation of the storage air, which keeps leaf tissues green and viable, hence less susceptible to attack. Transit temperatures maintained around 4.4 °C and storage at 0 to 0.6 °C but not above 2 °C are required for reducing postharvest losses due to this disease and maintaining good quality.

Field sanitation and preharvest sprays of copper oxychloride have been found to be the most effective method to control decay. Storage diseases also can be controlled by pre/postharvest application of fungicide such as iprodione, 50 per cent WP at a dose of 67 g/100 liters of water. The heads should be packed in perforated polyethylene bags and stored immediately. Treated cabbages must not be removed from store for at least eight weeks after treatment.

3.3 Carrot

The most common storage diseases in carrot are black mould, black rot and sour rot. The diseases produce discolouration and the affected tissues become soft or mealy. With sour rot fermentative odour is also produced. The Black mould/root rot is caused by *Thielaviopsis basicola*, black rot by *Stemphyllium* sp., greenish black mould by *Alternaria radicina*, sour rot by *Geotrichum candidum*, crown by *Rhizoctonia solani* and dry rot by *Rhizoctonia* sp. In cold storage rots due to *Rhizopus* sp. and cottony soft rot due to *Sclerotinia sclerotiorum* are problems and usually appear after some months of storage. At higher temperature, *Fusarium* sp. and *Rhizopus* sp. attack the roots. *Sclerotium rolfsii* is another pathogen which causes heavy yield loss due to the rotting of the roots. Bacterial soft rot is caused by the bacterium *Pectobacterium carotovorum* (*Erwinia carotovora*). Affected carrots develop soft lesions and eventually become a mushy and slimy mass of tissue. Two new postharvest pathogens, *Mycocentrospora acerina* and *Fibularhizoctonia carotae*, which cause liquorice rot and crater rot respectively, during prolonged low temperature storage of carrots have been recorded in Norway. Some postharvest rots of carrot are shown in Figure 13.

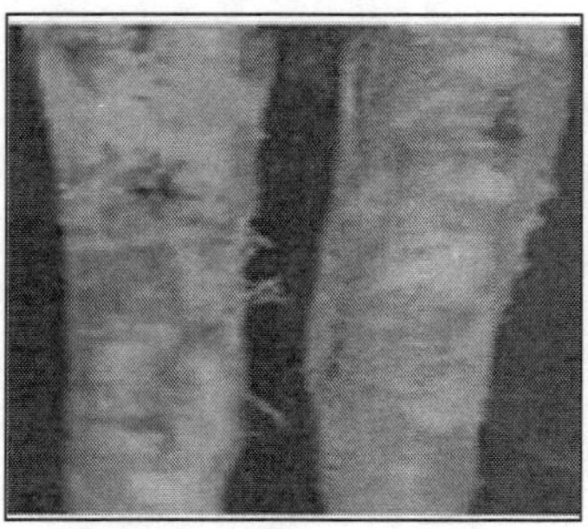

(a) *Alternaria* Rot
(http://agritech.tnau.ac.in)

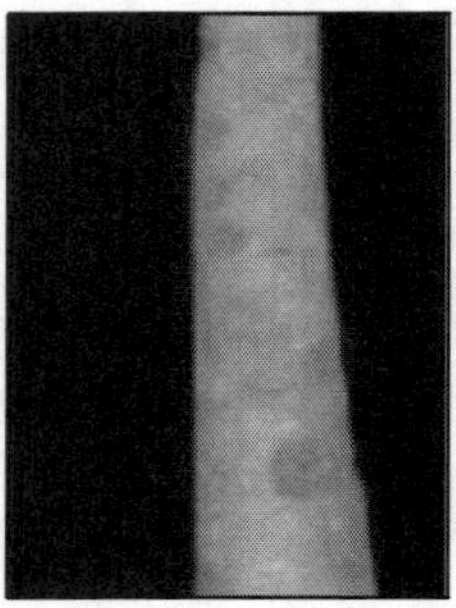

(b) Bacterial Soft Rot
(https://www.agric.wa.gov.au)

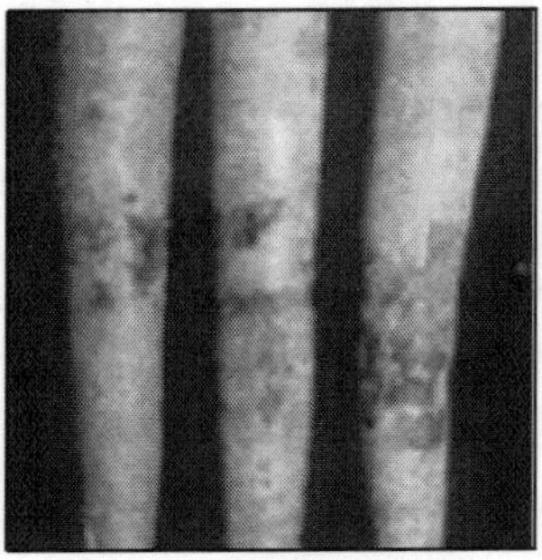

(c) Black Root Rot
(https://extension.umass.edu)

(d) Cottony Soft Rot
(http://agritech.tnau.ac.in)

(e) Crown Rot
(http://agritech.tnau.ac.in)

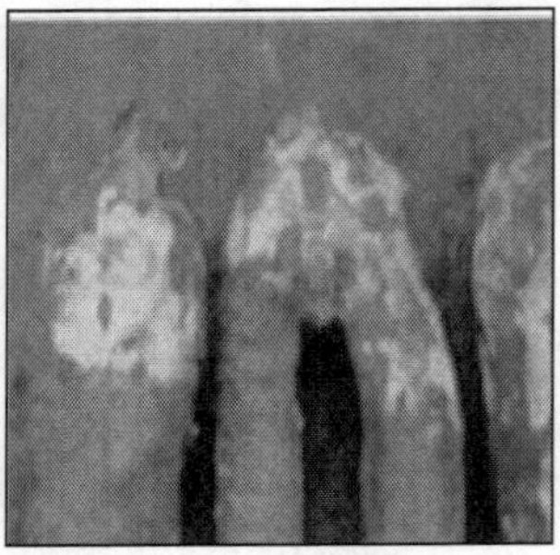

(f) Dry Rot
(http://agritech.tnau.ac.in)

Figure 13: Postharvest Diseases of Carrot

Handling of carrots to avoid mechanical injury is the best control measure. Postharvest washing is in chlorinated water and storage temperature of 0 °C is recommended.

3.4 Onion

3.4.1 Diseases

Onion is one of the few vegetable crops that can withstand rough handling, distant transportation and storage to an extensive degree. Apart from onion diseases discussed below, other fungi associated with postharvest rotting are *Fusarium oxysporum*, *Penicillium ducaleeni* and *Rhizopus stolonifier*. Estimated storage losses of up to 60 per cent in varieties like Nasik-Red, Bellary-Red and Pune Selection could be caused due to various fungi like *Sclerotium rolfsii*, *Aspergillus niger*, *Alternaria Solani*, *A. cepulicola* and *Fusarium* sp. (Figure 14)

a) Black Mould Rot

In this disease, black powdery spores of the fungus adhere to the exterior scale or in between the outer two layers. The disease is most common in dry conditions and could be very destructive in storage. The main organism involved in this disease is *Aspergillus niger* causing postharvest losses up to 20 per cent (Figure 14a).

b) Bulb Rot

This disease is most common in warm regions and is mainly due to *Fusarium solani*. The base of the bulb is usually the first site of infection. Infected onions become soft and pale-white in colour. In 7-8 days rotting completely consumes the bulb. The rate of decay is very slow below 15 °C (Figure 14b).

c) *Cephalosporium* Rot

Cephalosporium curtipes when inoculated in bulbs, caused 26.6 percent rotting but natural incidence in India is not of economic importance.

d) Neck Rot

The disease is caused by *Botrytis allii* Munn. The fungus causes softening of scales, which become soaked in appearance, later greyish, sporulating mycelial mat develops on the surface of the scales. The disease is severe when moist conditions prevail just before and during harvest and while onions are cured in the field (Figure 14c).

e) Purple Blotch

Alternaria porri (E11) Cif. Infection with this pathogen causes severe storage loss with development of purple blotch. But one thing is established that in the field the disease is initiated with the infestation of onion thrips, *Thrips tabaci*. In storage the infection becomes more serious (Figure 14d).

f) Soft Rot

Whole onion appears to be water soaked and colour of individual fleshy scale varies from yellow to light brown in this disease. The diseased scales are soft. The micro-organisms involved are species of *Erwinia* and *Pseudomonas* (Figure 14e).

3.4.2 Disease Management

Postharvest losses in onion can be reduced by proper curing and storing at low temperatures 0 to 2 °C and moderate relative humidity 65-75 per cent under appropriate conditions. Ventilating the onions in special drying chambers with air at 30 °C to reduce the moisture content and then treating them for 8-12 hours at 45 °C before transferring to the store, takes care of several onion pathogens in store.

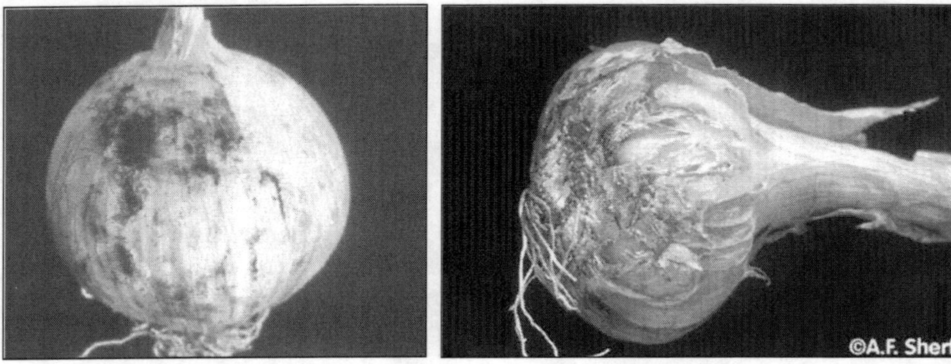

(a) Black Mould Rot
(http://mtvernon.wsu.edu/path_team)

(b) Bulb Rot
(http://vegetablemdonline.ppath.cornell.edu/)

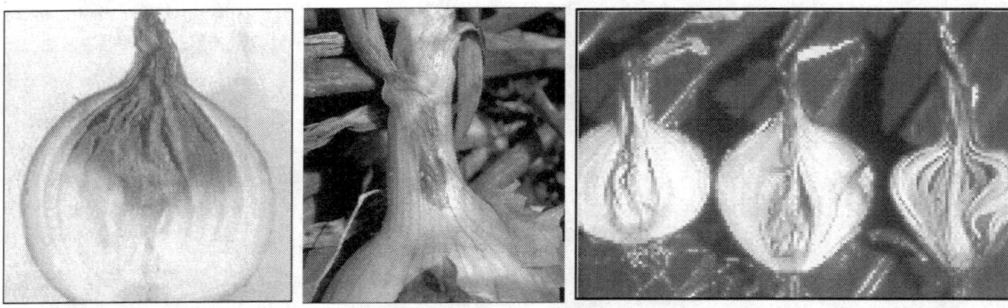

(c) Neck Rot
(https://www.rhs.org.uk and http://omafra.gov.on.ca)

(d) Purple Blotch
(http://gardener.wikia.com/)

(e) Soft Rot
(http://mtvernon.wsu.edu/path_team)

Figure 14: Postharvest Diseases of Onion

Preharvest sprays of tricyclazole or zineb in the field control purple blotch in the storage. National Horticultural Research and Development Foundation, Nashik, has recommended treating of bulbs with acetic acid which controls black mould very efficiently. Preinoculation treatments with 2,4-D, potassium metabisulphate, diphenylanine and acetic acid gave 58 per cent reduction in rotting due to *Fusarium* bulb rot, but none of the post inoculation treatments was effective in controlling the disease.

A formulated product of *Coniothyrium minitans* Campbell, biocontrol agent for soil borne pathogens is available in other countries to control onion basal and neck rots.

3.5 Potato

3.5.1 Diseases

In potato an average loss of 22.8 per cent is due to fungal diseases during storage and marketing. Apart from important postharvest diseases (Figure 15) discussed in subsequent paragraphs, some other important storage diseases that have been reported in potato are: dry rot of tubers due to *Aspergillus niger*, tuber rot due to *Penicillium herquei* and *Pythium aphanidermatum* and black leg due to bacteria *Erwinia caratovora*.

a) Black Sour Rot

Black sour or *Rhizoctonia* rot of potato is caused by its perfect stage *Corticium solani*. Black sclerotial bodies are formed on the diseased tubers. The fungus infects the stem and tubers start rotting in the field which continue in storage also.

b) Brown Rot

The brown rot causing pathogen (*Pseudomonas solanacearum*) survives on tuber surface for months. The tubers harvested from bacterial wilt affected fields carry infection. The infected tubers show ring of water soaked area in vascular region on cutting, which later turns dark brown to black. Creamy bacterial ooze can be seen oozing from vascular regions in severe infections. Dry cracks can be seen on tuber surface. Isolations carried out from rotten tubers collected from different places yielded species of *Erwinia, Bacillus* and *Pseudomonas* as well. The incidence of bacterial rot increases in late blight infected and late stored tubers (Figure 15a).

c) Charcoal Rot

The fungus *Macrophomina phaseolina* is the cause of this rot in warm dry periods, leading to heavy losses in different periods and places. The lenticels appear black and the flesh underlying skin also becomes uniformly black up to 2-5 mm depth. Secondary rot causing organisms lead to soft rot, otherwise dry rot with cavities in invaded tissues can be observed in later stages (Figure 15b).

d) *Cylindrocarpon* Dry Rot

Cylindrocarpon tonkinensis causes dry rot of tuber in storage under normal conditions. It appears as small brown patches on the surface of tubers having extensive white mycelial growth. Internally the flesh becomes dirty grey. The tubers undergo wet rot if the humidity exceeds 60 per cent resulting in pulpy decay with thick brown exudate full of mycelium and spores. In storage the rot may progress until the affected potatoes become quite hard and mummified and are completely destroyed.

e) Dry Rot

The disease develops in stores after few months of harvest and is of worldwide occurrence. The pathogens *Fusarium solani* var *coerulum*, enter through proliferated lenticels, injuries or scab pustules. The infected skin becomes dark in colour and as it spreads, the lesions become sunken. The sunken skin becomes wrinkled, which are prominent and arranged in concentric circles around point of infection. The tubers become dry, hard, shriveled and light weight. Inside cavities with fungal mycelium and blue stained wall are observed. The susceptibility of tubers to dry rot increased with the storage period at room temperature and the maximum infection was observed after 4 months of storage in country stores. Among the commercial varieties, the incidence of dry rot was maximum in Kufri Bahar. Some other Fusarium species are also reported to cause dry rot (Figure 15c).

f) *Gilmanella* Rot

Disease caused by *Gilmanella humicola* Bavun has been reported on late blight differential varieties of potatoes in storage. Minute brownish necrotic spots around lenticels and eyes were observed during storage at room temperature (15 °C to 25 °C). Almost all the eyes of affected tubers become shriveled. Affected cells become brown, thick walled and crushed. Such tubers are unfit for seed purpose.

g) Late Blight

This disease originates in the field. The tubers start rotting during storage and marketing. Saprophytic bacteria and other fungi are often associated with late blight lesions, leading to further deterioration. The fungus *Phytophtora infestans* infects the tuber still lodged in soil causing dry rot and brown discolouration of the tissues. In wet soil the tubers may rot quickly due to opportunistic infection of soft rot bacteria. Potato variety Kufri Jyoti is resistant to this disease (Figure 15d).

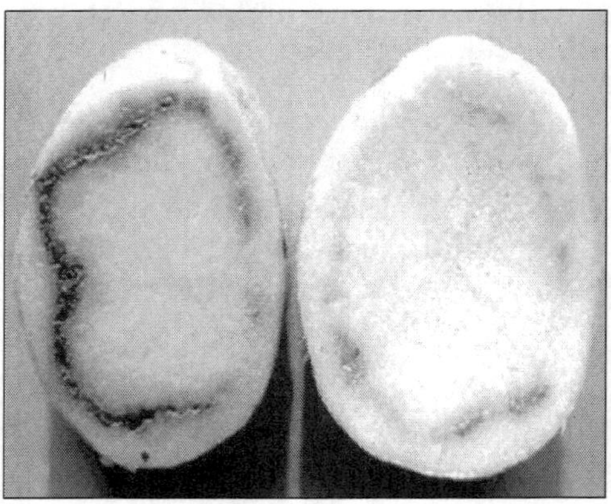

(a) Brown Rot
(http://www.cals.ncsu.edu and http://www.agf.gov.bc.ca/cropprot/brr.htm)

(b) Charcoal Rot

(c) Dry Rot

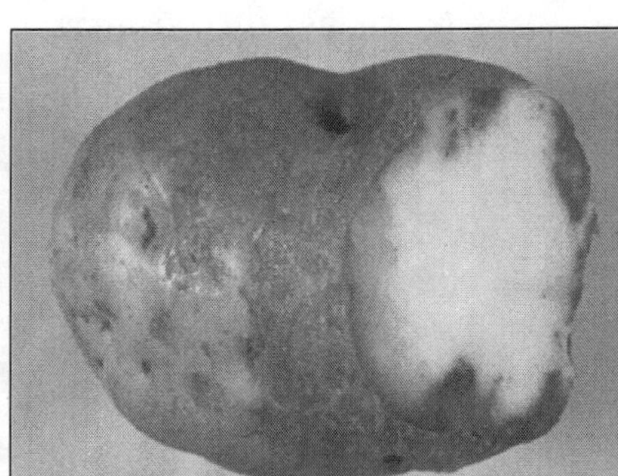

(d) Late Blight
(http://archivebio.ed.ac.uk)

Figure 15: Postharvest Diseases of Potato

h) *Rhizopus* Rot

It is a damaging postharvest disease resulting in soft rot of tubers. *Rhizopus oryzae* and *R. arrhizus* infections are responsible for the decay of stored potato tubers. In case of high infection, losses can reach up to 100 per cent during transit storage.

3.5.2 Disease Management

Potato tubers are harvested from soil, which is an abode for several pathogenic as well as beneficial microflora. Hence, postharvest rots can be greatly reduced by manipulating the soil ecosystem to be more suppressive. General field sanitation, crop rotation and green manuring greatly influence soil and tuber borne diseases by increasing beneficial microflora and improving plant nutrition.

Killing or drying of haulms for a month before harvest and pre storage curing for 15 days is important to reduce storage rot. During harvesting, care should be taken to avoid injuries and damaged tubers should be sorted out before storage.

Potato tubers can be stored at 5-10 °C to reduce rots, however, for table purpose the tubers turn sweet at low temperature. Hence storage in non-refrigerated stores (16-30 °C with 75-95 per cent RH along with chemical treatment for sprout suppressants (CIPC, MH etc) for 3-4 months, till rains start is recommended.

Boric acid (4 per cent), reduced the rot incidence to 7.4, as compared to 16.6 per cent in control (water) when potatoes were stored for 138 days at average minimum and a maximum fortnightly temperature of 28 °C and 32.3 °C at humidity of 67.5 and 78.5 per cent respectively. Disease severity of *Rhizopus* rot was significantly reduced by calcium chloride and fluorescent pseudomonads used alone or in combination, and it was nearly suppressed by azoxystrobin.

3.6 Tomato

3.6.1 Diseases

Tomato fruit is the most widely cultivated vegetable, available in all seasons and cherished throughout the world. A large number of fungi and bacteria infect tomato that result in postharvest diseases (Figure 16), which reduce quantity as well as quality. The fruit is highly perishable and most of the infections of tomato occur during unripe stage in field, because almost all postharvest pathogens are also present on foliage parts in field. The infections later develop during postharvest handling and storage. Management throughout the season is therefore, critical for a quality produce and profit.

a) *Alternaria* Rot

It is a field as well as storage disease. The fungus - *Alternaria alternata* enters through stem end or injuries and results in black depressed spots with concentric rings around the point of infection. Flesh is also affected (Figure 16a).

b) Anthracnose

Black dot disease in tomato is very common and is caused by different species of *Colletotrichum e.g. Colletotrichum phomoides, C. coccodes, C. dematium* and

C. gloeosporioides. Lesions are small circular, restricted and depressed. They may enlarge to 12 mm or more diameters. The lesions are zonate and surface may appear salmon-coloured (Figure 16b).

c) Bacterial Soft Rot

Bacterial soft rot of tomato is caused by *Erwinia carotovora* Holland, *E. atroseptica* Jenninson, E. *aroideae* Holland. It causes extensive loss in storage. The bacteria infect the fruits and produce water soaked lesions which enlarge rapidly by making the tissue soft and discoloured. Ultimately the affected tissue disintegrates into a mass of disorganised cells. In controlled atmosphere with 3 per cent O_2 and 5 per cent CO_2, the disease incidence is low in inoculated tomatoes (Figure 16c).

d) Bacterial Spot

The bacterial spot of tomato is caused by *Xanthomonas vasicatoria*. When the relative humidity exceeds 85 per cent and the average temperature is 20 °C, the disease spreads rapidly. The bacterial spots generally affect green fruits but sometimes these are also observed on mature fruits. The size of the spot may be 6 mm in diameter. Spots on fruit begin as dark raised spots surrounded by a water-soaked border. As the fruit expands, the spots may appear to be sunken in pits, as the healthy tissue will grow faster than the spotted tissue. Older spots may reach a diameter of 1/3" and are black, slightly raised and have lobed borders. They do not penetrate the seed cavity of the fruit, but are superficial. As they age, the centers of the spots disintegrate and sink, and eventually form a crater. The skin of the fruit is often cracked at the edge of older spots. The water soaked border is absent in older spots (Figure 16d).

e) Buck Eye Rot

This disease is caused by *Phytophthora nicotianae* which is soil and seed borne and favoured by high temperature and humidity. It is more severe in poorly drained soils. On fruits, the symptoms are greyish green or brown water soaked spots on the fruits near soil level. Lesions enlarge on the surface of infected fruit with characteristic pattern of alternating light and dark brown concentric rings which give this disease its name of buckeye. Buckeye rot lesions have a smooth surface and lack a sharply defined margin. The rot remains firm unless invaded by secondary rot causing organism (Figure 16e).

f) *Cladosporium* Rot

Dry rot of the fruit pericarp region is caused by a species of *Cladosporium* and the spread of the disease is favoured during rainy weather. The rot is confined to pericarp region and does not penetrate the pulp.

g) *Fusarium* Rot

The disease is caused by *Fusarium roseum* Link. and is characterised by the appearance of water soaked lesions on ripe fruits. The fungus sporulates over the diseased area. Later the decay progresses rapidly causing soft disintegration of the pulp (Figure 16f).

h) Grey Mould

Grey mould caused by the fungus *Botrytis cinerea*, an amorphic form of *Botryotinia fuckeliana*, represents one of the main causes of decay in harvested tomatoes. The most common symptom is dirty white growth of the fungus on the lesion, which are soft water soaked. In varieties where infection of fungus is unable to develop further, 3 to 8 mm in diameter faint halo appear on fruits. These halos are white on immature fruits and yellow on ripe fruits. Spots do not develop further, though a small necrotic speck may appear in the spot (Figure 16g).

i) *Malustela* Rot

The disease is caused by *Malustella seria* Batista. Irregular lesions produced on the ripe fruits are surrounded by the black spore masses of the fungus. The rot produced by the fungus is firm and not accompanied by any off-smell. The external symptom is similar to *Alternaria* rot.

j) *Oospora* Rot

This rot is also known as watery rot and is caused by *Oospora lactis*. The fungus attacks the stem end or the growth crack and invades the cellulose walls of the fruit. The decayed tissues are covered with white velvety or granular fungal growth. The affected tissues emit unpleasant odour.

k) *Phoma* Rot

Rot due to *Phoma destructiva* Plowr causes heavy losses in transit and storage. Lesions appear on wound or insect puncture sites or stem end scar cracks. The spots on fruit are sunken black of increasing size. The spots develop black pycnidia in later stages (Figure 16h).

l) *Rhizopus* Rot

The disease caused by *Rhizopus nigricans* Ebr. causes severe loss of fruits in transit and storage. The fungus attacks the ripe fruits. The affected fruits become soft and watery exuding clear watery fluids. The fungus penetrates deep into the tissue of the pulp and causes a soft rot covered with thick cottony mycelial growth having black sporulation. Favourable temperature for rot development is 18-24 °C and RH a) above 50 per cent (Figure 16i).

m) *Sclerotium* Rot

This is caused by *Sclerotinia sclerotiorum* (Lib.) de Bery. The spots are slightly sunken, yellow on diseased fruit, which decay rapidly and infected tissues collapse. The lesion is covered by a white fungal mass dotted with numerous black sclerotia. One infected fruit can destroy the entire lot (Figure 16j).

n) Sour Rot

It is caused by *Geotrichum candidum* and is the most serious disease of tomato grown in the monsoon season. This is characterized as water soaked lesions followed by the cracks in the fruit skin which ooze out watery excretion. On the cracks white fluffy tuft of the fungus is prominent. If not controlled, the entire lot gets infected overnight emitting characteristic foul smell (Figure 16k).

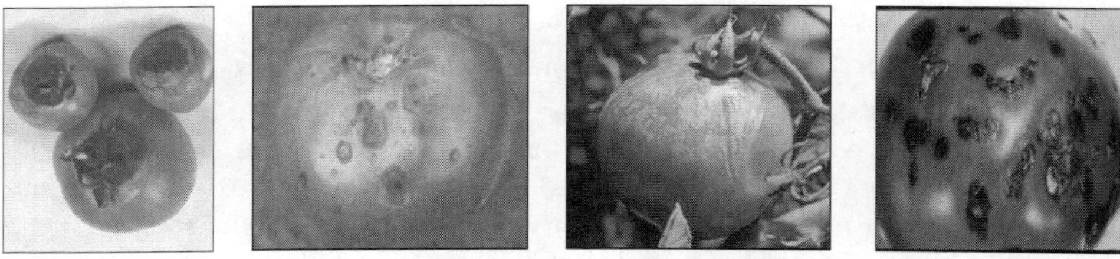

(a) *Alternaria* Rot **(b) *Anthracnose*** **(c) Bacterial Soft Rot** (http://agritech.tnau.ac.in) **(d) Bacterial Spot** (http://ipm.uconn.edu)

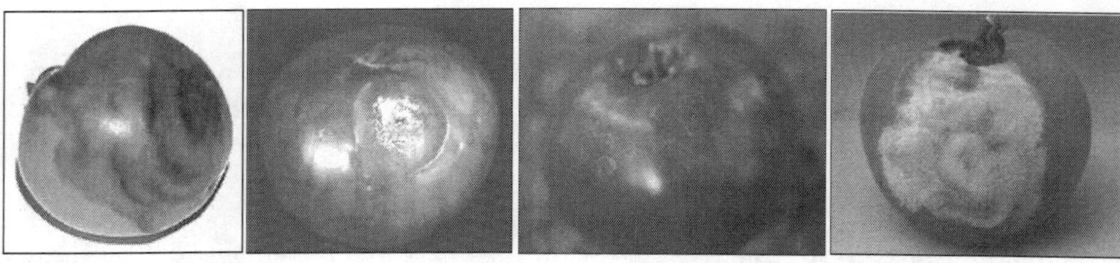

(e) Buck Eye Rot (http://pubs.ext.vt.edu) **(f) *Fusarium* Rot** (http://agritech.tnau.ac.in) **(g) Grey Mould Rot** (http://pnwhandbooks.org and www.interempresas.net)

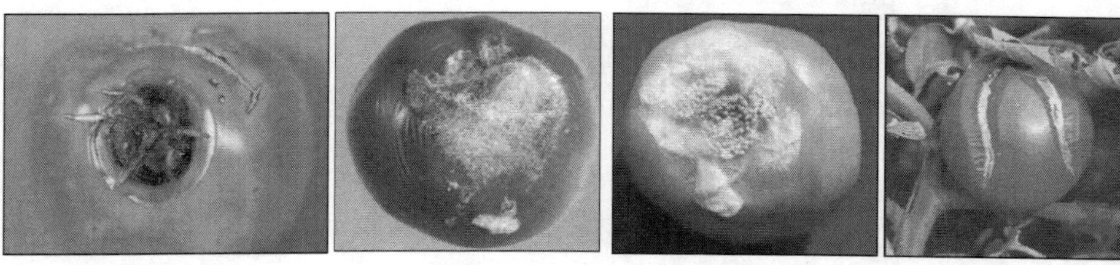

(h) *Phoma* Rot (http://erec.ifas.ufl.edu) **(i) *Rhizopus* Rot** (http://agritech.tnau.ac.in) **(j) *Sclerotium* Rot** (http://agritech.tnau.ac.in) **(k) Sour Rot** (http://allaboutgrowingtomatoes.blogspot.in)

Figure 16: Postharvest Fungal Diseases of Tomato

3.6.2 Disease Management

Tomato being very succulent and juicy is very delicate to handle. Cultural practices like staking the plants and removing the foliage and fruits up to 30 cm from the ground level have been effective in reducing various rots. This enables sunshine and air to circulate and improve microclimate. Sanitation and cultural practices are very important for the control of soft rots. Debris should be removed from the warehouses each season before disinfecting for new storage. Before storage, fruits with scars, cracks, injuries or soft fruits should be sorted out. In controlled atmosphere with 3 per cent O_2 and 5 per cent CO_2, the disease incidence is low. The tomato fruits can be stored for prolonged periods at 7 to 12 °C along with fungicidal treatments. Red fruits are better protected than green at low temperatures.

When the fruits are transported to long distance, they should be dipped in 0.2 per cent mancozeb. Effective check of rots is also achieved with chlorothalonil treatments. Treatment with Thiobendazole 1000 ppm has been found to be effective in controlling the storage disease of tomato caused by *Colletotrichum dematium*. Spraying of streptomycin (Agrimycin 100) @1000 ppm or Bavistin (0.2 per cent) at 10 days interval provides good control of bacterial diseases.

A number of bacterial and fungal antagonists have been found to effectively control postharvest rots of tomatoes. They have various modes of action that include antibiosis and/or competition for nutrients and space. This method of control for postharvest rots of vegetables appears to be feasible and may present an alternative to synthetic pesticides. Application of *Trichoderma harzianum* in the field and also as a dip before storage has been recorded to be quite effective.

4.0 Conclusion

Postharvest diseases have both nutritional as well as economical significance and the problem is more complex when dealing with management of fruits and vegetables, which are more perishable than other crops. The knowledge and application methods adopted to reduce postharvest losses are more advanced in developed countries. The developing nations have vast knowledge but are slow to understand and adopt technologies for handling postharvest produce to get high prices.

Integrated management practices should be designed taking into consideration the storage life of the commodity and the extent and type of storage losses the commodity can face. The management methods should include means to avoid the entry of pathogens, through various field operations before harvest, build up resistances in plant and keep pathogen load below economic threshold. Handling methods should be safe so as not to damage the produce, restricting entry of most of the storage pathogens which are weak parasites. Physical methods should be followed to avoid the pathogen damage by storing produce under proper storage conditions suitable for the commodity and using suitable packing, wrapping materials *etc.* Methods to treat the produce using preharvest chemical or biological sprays or postharvest dips, growth regulators, hot water treatment, radiation therapy, chemical/ biological wraps should also be followed. The management strategy should be such as to give maximum returns to the investment made.

Safety of consumers should be our first priority. Use of chemicals should be done taking into consideration the preharvest interval of the chemical and its registration for intended use. Biocontrol organisms also need to be used with precaution and complete knowledge as some of the bioagents like *Burkholderia cepacia* have now been established to be an opportunistic human pathogen. *Pantoea agglomerans* once widely used as bioagent is now banned because it is known to cause skin irritation. In fact some pathogens produce mycotoxins which can harm human beings and animals. Hence, the methodology used should be best suited to the crop with maximum benefits and human safety.

5.0 Important References

Agarwala, R. K. and Sharma, V. C. 1968. Storage rot diseases of apple. *Indian Phytopath.*, **21**: 294-298.

Aulakh, K. S. and Grover, R. K. 1968. Ripe fruit rots of tomato and their control by oils. *Pl. Dis. Rep.*, **52**: 355 – 359.

Baraldi, E., Mari, E., Chierici, E., Pondrelli, M., Bertolini, P. and Pratella, G. C. 2003. Studies on thiabendazole resistance of *Penicillium expansum* of pears: Pathogenic fitness and genetic characterization. *Plant Path.*, **52**: 362–370.

Batta, Y. A. 2007. Control of postharvest diseases of fruit with an invert emulsion formulation of *Trichoderma harzianum* Rifai. *Postharvest Biol. Tech.*, **43**: 143–150.

Bhargava, S. N. and Singh, A. P. 1975. *Aspergillus* rot of mango and its control. *Indian J. Hort.*, **32**: 190.

Chakravarti, D. K. and Srivastava, D. N. 1964. Stem end rot of mango and orange fruits incited by *Diplodia natalensis* Pole Evens. *Curr. Sci.*, **33**:285.

Chona, B. L. 1933. Preliminary investigations on the diseases of banana occurring in the Punjab and their method of control. *Indian J. agri. Sci.*, **3**: 673-687.

Costa, D. M. de. and Erabadupitiya, H. R. U. T. 2005. An integrated method to control postharvest diseases of banana using a member of the *Burkholderia cepacia* complex. *Postharvest Biol. Tech.*, **36:** 31–39.

Dang, J. K. and Gupta, P. C. 1985. Chemical control of storage rot of onion bulbs caused by *Fusarium solani*. *Haryana Agri. Univ. J. of Res.*, 1984, 403-404.

Dhingra, O. D. and Khare, M. N. 1971. A new fruit rot of papaya. *Curr. Sci.*, **40**: 612-613.

Dhingra, O. D., Khare, M. N. and Chand, J. N. 1970. Ciger-end disease of banana in India. *Curr. Sci.*, **39**: 498-499.

Dikbas, Neslihan and Bostan, Hidayet. 2009. Biological control of postharvest disease caused by *Aspergillus flavus* on stored lemon fruits. *African J. Biotech.*, **8:** 209-214.

Droby, Samir. 1991. Biological control of postharvest diseases of fruits and vegetables: alternatives to synthetic fungicides Review Article. *Crop Prot.*, **10**: 172-177.

Droby, S., Porat, R., Cohen, L., Weiss, B., Shapira, B., Philosoph-Hadas, S. and Meir, S. 1999. Suppressing green mould decay in grape fruit with postharvest jasmonates application. *J. Am. Soc. Hort. Sci.*, **124:** 184–188.

Fravel, D. R. 2005. Commercialization and implementation of biocontrol. *Ann. Rev. Phytopath.*, **43**: 337–359.

Govender, V., Korsten, L. and Sivakumar, D. 2005. Semi-commercial evaluation of *Bacillus licheniformis* to control mango postharvest diseases in South Africa. *Postharvest Bio and Tech.*, **38** : 57–65

Grover, R. K. 1965. *Phoma* rot of tomato *Indian Phytopath.*, **18**: 388-389.

Gupta, R. P. and Srivastav, P. K. 1984. Study on storage diseases of onion. *Indian Phytopath.*, **37**: 404.

Hermansen, A., Wanner, L. Nærstad, R and Klemsdal, S. S. 2012. Detection and prediction of postharvest carrot diseases. *Eu. J Pl.Path.*, **133**: 211-228.

Kapoor, I. J. and Tandon, R. N. 1970. A new species of *Macrophoma* causing fruit rot of guava (*Psidium guajava*). *Indian Phytopath.*, **23**: 122-125.

Kapur, S. P. and Chohan, J. S. 1974. Factors affecting infection and development of fruit rot in papaya by *Macrophomina phaseoli. Indian Phytopath.*, **27**: 252-253.

Kaushik, C. D., Thakur, D. P. and Chand, J. N. 1972. Parasitism and control of *Pestalotia psidii* causing cankerous disease of ripe guava fruits. *Indian Phytopath.*, **25**: 61-64.

Kefialew, Y. and Ayalew, A. 2008. Postharvest biological control of anthracnose (*Colletotrichum gloeosporioides*) on mango (*Mangifera indica*). *Postharvest Biol. Tech.*, **50**: 8–11.

Khanna, K. K. and Chandra, S. 1966. Control of storage diseases of fruits. *Curr. Sci.*, **45**: 70.

Khanna, K. K. and Chandra. S. 1974. A new fruit rot of banana in India. *Curr. Sci.*, **43**: 491-492.

Khanna, R. N. and Sharma, Jyotsana. 1993. Soil and tuber borne diseases. In: *Advances in Horticulture. Vol. 7- Potato*. K. L. Chadha and J. S. Grewal. (Eds.), Malhotra Publishing House, New Delhi, pp. 463-490.

Kortekamp, A. 2006. Effectiveness of calcium salts, hydrogen peroxide, azoxystrobin, and antagonistic bacteria to control postharvest rot on tobacco caused by *Rhizopus oryzae*. *Int. J Pest Manage.*, **52**:109-115.

Kothari, K. L. 1968. Controlling fruit rot of guava. *Indian Hort.*, **12**: 9-10.

Lahlali, R., Serrhini, M.N. and Jijakli, M.H. 2005. Development of a biological control method against postharvest diseases of citrus fruit. *Comm. Agri. Appl. Biol. Sci.*, **70**: 47–58.

Lingaraj, D. S. 1969. Mango diseases and pests. *J. Mysore Hort. Soc.*, **14**: 15-18.

Long, C. A., Deng, B.X. and Deng, X. X. 2007. Commercial testing of *Kloeckera apiculata*, isolate 34–9, for biological control of postharvest diseases of citrus fruit. *Ann. Microbiol.*, 57: 203–207.

Mehta, P. R. 1940. Stem end rot and soft rot of pineapple in the United Province. *Curr. Sci.*, **9**: 330.

Nayak, M. L. 1964. A new organism causing dry rot of potatoes. *Sci. Cult.*, **23**: 605-606.

Patel, M. K., Kamat, M. N. and Hingorani, G. M. 1950. *Pestalotia psidii* on guava. *Indian Phytopath.*, **24**: 115-118.

Prasad, B. G. R. 1978. An analysis of problems of marketing of few fresh vegetables in Bangalore city. M.Sc.(Agri.). Thesis, Univ.Agri. Sci. Bangalore.

Prasad, S. S. and Verma, R. A. B. 1970. Fungal diseases of papaya fruits in Bihar. *Indian Phytopath.*, **23**: 722-724.

Rai, J. N. 1956. Styler end rot of guava fruit (*Psidium guajava* L.). *Proc. Indian Acad. Sci.*, **43**: 55-61.

Rawal, R. D., Muniyappa, N. C. and Ullasa, B. A. 1983. Effect of preharvest field sprays of the fungicides on the control of storage rot of papaya. *Indian J. Agri. Sci.*, **53**: 614-615.

Reddy, S. N. and Reddy, C.K.R. 1983. *Greenaria* fruit rot: an endemic diseases of grapes in India. *Indian Phytopath.*, **36**: 110-114.

Sharma, Jyotsana, Sharma, K. K. and Jadhav, V. T. 2012. Diseases of Pomegranate. In: *Diseases of Fruit Crops* A. K. Misra, P. Chowdappa, Pratibha Sharma and, R. K. Khetrapal (Eds.), Indian Phytopathological Society, New Delhi: pp. 181-224.

Sharma, R. R., Singh, D. and Singh, R. 2009. Biological control of postharvest diseases of fruits and vegetables by microbial antagonists: A review. *Biol. Cont.*, **50**: 205–221.

Singh, D. 2002. Bioefficacy of *Debaryomyces hansenii* on the incidence and growth of *Penicillium italicum* on Kinnow fruit in combination with oil and wax emulsions. *Annals of Pl. Prot. Sci.*, **10**: 272–276.

Spadaro, D., Garibaldi, A. and Gullino, M.L. 2004. Control of *Penicillium expansum* and *Botrytis cinerea* on apple combining a biocontrol agent with hot water dipping and acibenzolar-S-methyl, baking soda, or ethanol application. *Postharvest Biol. Tech.*, **33**: 141–151.

Srivastava, M. P. and Tandon, R. N. 1969. Postharvest diseases of guava in India. *Plant Dis. Rept.*, **53**: 206-208.

Srivastava, M. P. and Tandon, R. N. 1971. Postharvest diseases of papaya. *Int. J.Pest Manage.*, **17**: 51-64.

Srivastava, M. P. and Tandon, R. N. 1971. Postharvest diseases of banana in India. *Indian Phytopath.*, **24**: 115-118.

Tandon, R. N. and Mishra, A. N. 1969. Fruit rot diseases of *Carica papaya* and *Musa paradisia* caused by *Rhizopus Stolonifier*. *Indian Phytopath.*, **22**: 334-341.

Vir, D. and Gaur, A. 1971. Postharvest chemical treatment of apple fruit against rot and deterioration. *Curr. Sci.*, **40**: 387.

Xi, L. and Tian, S.P. 2005. Control of postharvest diseases of tomato fruit by combining antagonistic yeast with sodium bicarbonate. *Sci. Agri. Sinica.*, **38**: 950–955.

Yu, T., Zhang, H., Li, X., and Zheng, X. 2008. Biocontrol of *Botrytis cinerea* in apple fruit by *Cryptococcus laurentii* and indole-3-acetic acid. *Bio. Cont.*, **46**: 171–177.

Zhou, T., Chu, C.L., Liu, W.T. and Schneider, K.E. 2001. Postharvest control of blue mould and grey mould on apples using isolates of *Pseudomonas syringae*. *Can. J. Pl. Pathol.*, **23**: 246–252.

Zhou, T., Northover, J., Schneider, K.E. and Lu, X.W. 2002. Interactions between *Pseudomonas syringae* MA-4 and cyprodinil in the control of blue mould and grey mould of apples. *Can. J. Pl. Pathol.*, **24**: 154–161.

Chapter 7

Grading and Packaging

M.S. Ladaniya

*Director, ICAR-National Research Centre for Citrus,
Amaravati Road, Shankar Nagar Post Office,
Kachimet, Nagpur – 440 010, Maharashtra, India
E-mail: msladaniya@gmail.com*

1.0 Introduction

Grading and packaging of fresh fruits is one of the most important steps in the long and complicated journey from grower to consumer and has direct bearing on price realized in the market. These are two important links in entire supply chain from grower to consumer. Plastic crates, baskets, cartons, wooden boxes, bulk bins, and palletized containers are the common containers for handling and transporting fresh fruits.

While grading helps in enhancing value of the fruit, proper packaging can reduce bruising, crushing and moisture loss, prevent re-contamination of the product with spoilage organisms, reduce pilferage and maintain a sanitary environment during marketing and thus improves sale of produce.

2.0 Grading

2.1 Importance

After receiving and preparing fruit in packhouse, grading is next important operation. Fruit grade indicates and assure internal and external quality to the buyer and fetches good price for the producer. Before grading, sorting of fruit is necessary. It helps in removing culled fruits of all types including too large or too small and over ripe and under-ripe fruits. Sorting and grading in large operations are mechanized.

Sorting and presizing operations are done at the same time in advanced systems where very large and very small fruits are also removed as culls along with blemished fruits. Commercial scale sorting conveyers made of aluminium rollers are used where trained people remove culled fruits manually. Machinery of various capacity (from 2 - 5 tonne per hour capacity with 60-150 cm width of the conveyer) is available depending on requirement. The rollers rotate on their axis in order to rotate the fruit to view entire surface while fruit moves forward. The culled fruit needs to be removed quickly from the area and hence such sorting conveyers need extra space around for fruit handling. In case of manual sorting the fruit sorting personnel have to be sufficiently trained to identify culls on the basis of colour, size, shape, blemishes, etc. The sorting conveyer should have sufficient light/illumination focused on conveyer and fruit. Supervisors take care for proper sorting to avoid losses (good fruit removed along with the culls) or any "under sorting" (culls being retained in good fruit). In advanced automatic electronic sorting systems, fruits can be sorted on the basis of colour, size, shape and blemishes by cameras (Machine–vision system) which can detect culls and drops and remove them on cross belts running below. These operations are possible with the help of image analysis by the computer which is programmed for the purpose. Fruits can be sorted in different colour categories as desired. For example - lemons can be sorted in 4 colour classes – dark green, light green, silver and yellow by electronic sorters.

2.2 Methods

Grading of fruits is done visually on the basis of surface blemishes and colour or any other deformity as the fruit rolls on the conveyer. Conventionally it is done manually by experienced graders. Graders have to be trained for recognizing percentage of blemished part on the fruit surface since fruits having 5-10 per cent surface blemishes are some times allowed for various grades. Most experienced graders should be at last point of the conveyer so that no off-grade fruit is included. Fruits of second grade are removed and packed later and off-grade fruit is sent for processing/ juice extraction.

Grading can also be done mechanically using high-speed electronic graders which recognize blemishes and colour using photo–optical grading system. Electro-mechanical grader used in Sunkist Growers Inc. facility in California, USA, uses x-ray scanner (50 Roentgen/min) to scan the fruits for frost damage, granulation and *Alternaria* core rot. Then the fruit is graded for colour blemishes and scars using electronic machine–vision graders. This information is combined (internal and external) and computer operated software system decides the grade and marketability of fruit and it is dropped on proper conveyer for further process. The system is commonly used for lemons and oranges. The efficiency is as high as 480 fruits per min in each line (Johnson, 1991) and there could be several such lines in a packing house.

Sorting and grading are the two most labour intensive activities. These operations engage 40-50 per cent workforce in the packinghouse and even then results are not very satisfactory. There is thus scientific as well as economic interest in developing automatic sorting and grading machines. The 'Videograding TM' is a commercial automatic grading system used in Italy. The reliability and costs are the major factors in adoption of these technologies. New sensors are being developed to evaluate internal quality like sugars and acidity along with colour vision and multiple cameras are being used to locate external defects so that quality in totality can be given to consumers (Blandini *et al.*, 1992).

2.3 Grade/Quality Standards

There are various agencies that have established grade and quality standards including packaging regulations for fruits and vegetables. The most important at international level is Codex Alimentarius Commission of the WHO/FAO. Every country also has its own standards for perishable produce on the basis of which domestic trade is organized. This Commission is established by the Food and Agriculture Organization of the United Nations (FAO) and the World Health Organization (WHO) with the purpose of protecting the health of consumers and ensuring fair practices in the food trade.

(The *Codex Alimentarius* actually means "Food Code" in Latin. The commission's work is to collect internationally adopted food standards, guidelines, codes of practice and other recommendations and evolve a common standards or Code.)

The Commission has published Codex standards for fresh fruits and vegetables to allow their wide use and understanding by governments, regulatory authorities, food industries, retailers, and consumers in the world (Codex Alimentarius Commission, 2007). These standards are also revised/amended time to time particularly after World Trade Organization (WTO) came into existence and need to be followed by Indian fruit producers/packers during export keeping in view guidelines of importing countries. These guidelines include provisions concerning fruit quality, classification/grade of fruit, fruit sizes, packaging and presentation and hygienic/sanitary conditions to be followed.

2.4 Indian Grade/Quality Standards (Agmark Standards and Regulations)

The Indian "Agmark" standards and the standards of the Codex Alimentarius Commission are similar to a large extent. The Agmark standards are established by the Directorate of Marketing and Inspection, Department of Agriculture and Co-operation, Ministry of Agriculture, Government of India. These standards are established under Agricultural Produce (Grading and Marking) Act, 1937 as amended in 1986. It confirms the quality control and the best hygienic condition of the food. The certification also marks the food standards keeping in mind the requirements of WTO and benefits all including the producer and the consumer. Under this certification, the third party guarantee to quality is certified. Under the above mentioned act, rules and Amendments in rules take place from time to time. In recent past these rules were amended in 2004, 2007 and 2012 and these come into force from the date of their Official Gazette notification.

These rules apply to commercial varieties of Fruits and Vegetables. Agricultural Marketing Adviser to the Government of India or his representative is the certificate issuing authority. Producer/packer or trader who wants to follow these rule have apply for this certification and if granted, have to follow all rules laid down therein. The "Authorised packer" means a person or a body of persons who have been granted/issued a certificate of authorisation to grade and mark Fruits and Vegetables in accordance with the grade standards and procedure prescribed under these rules. The "Certificate of Authorisation" is issued under the provisions of the General Grading and Marking Rules, 1988 authorising a person or a body of persons to grade and mark Fruits and Vegetables with the grade designation mark. The grade designation mark shall consist of "AGMARK insignia" consisting of a design incorporating the certificate of authorisation number, the word "AGMARK", name of commodity and grade designation.

For domestic trade, fruits and vegetables shall comply with the residue levels of heavy metals, pesticides, aflatoxin and other food safety parameters as specified in Prevention of Food Adulteration Rules, 1955 while for export, fruits and vegetables may be graded and marked as per buyer's requirements in that country provided the minimum quality/grade requirements specified are met.

2.5 Grade Designations

The Agmark grade standards for quality of fruit describe classification of fruit in three grades or classes depending on quality of fruit.

In general for all fruits, the Grade designation/classification is suggested as follows:

i) Extra class
ii) Class – I
iii) Class - II

Besides Codex Alimentarius Commission and Agmark, several other agencies have also published standards for quality of fruits, particularly for exports to certain countries, *viz.* Agricultural and Processed Food Products Export Development Authority, New Delhi (APEDA), National Horticulture Board (NHB) and Himachal Pradesh Horticultural Produce Marketing and Processing Corporation Ltd. (HPMC).

2.6 Advantages of Grading

Grading is beneficial to the growers, traders as well as to the consumers, as it provides common standards to all and enables farmers to get better price for their produce. It assists the producers and other intermediaries in preparing fresh produce for market with appropriate labeling and helps the consumers to get standard quality produce at fair price. It facilitates the consumer to compare the prices of different qualities of a produce in the market and assures the quality of the produce and also reduces the cost of the marketing and transportation. Produce of similar grade can be stored in bulk and market values are better understood. Commodities can be bought and sold without inspection, through e-trading. Grading also provides an authentic and scientific basis in promoting and managing the marketing system. It serves as a realistic and common basis for market intelligence and reporting. It facilitates the settlement of quality disputes between buyers and sellers.

3.0 Packaging

Packaging is indispensible for the commodity to be marketed. Either at wholesale or retail or both stages fruit has to be packed for its delivery.Different type of fruit need different types of packages depending on the nature of their physical, anatomical and physiological (mainly transpiration, respiration and ethylene production rate) characteristics and susceptibility to microbial decay. For example strawberries would need different type of packing than mango and papaya. Jackfruit would need totally different packaging. Temperature, relative humidity and ventilation also play a very important role in determining the postharvest life of packaged fruit. Thus, the packaging for fresh fruit need to be designed based on considerations as follows:

i) Providing ventilation for respiration and exchange of gases

ii) Protection against moisture/weight loss

iii) Slowing down of respiration rate, delay in ripening and decay

iv) Control of ethylene concentrations in the package

v) Protection against bruising and physical injury

vi) Protection against microbial contamination and deterioration

All aspects of packaging must be taken into account when considering the introduction of new packaging. These aspects involve, among others, cost of packaging material, labour, acceptance by trader and consumers and changes in product condition. The ultimate goal of grading and packaging is a better quality to easier handling of the produce, and better marketable product.

3.1 Agmark Standards and Regulations for Packaging of Fruits

Agmark rules also prescribe standards for packaging of fruits. As per these rules fruits shall be packed in each container in compliance with the instructions issued by the agricultural marketing adviser, Government of India, from time to time for domestic market and the recommended International Code of Practice for Packaging and Transport of Fresh Fruit and Vegetables (CAC/RCP 44-1995, Amd. 1-2004) for export.

The general common regulations for all fruits are:

i) Fruits shall be packed in such a way as to protect the produce properly.

ii) The materials used inside the package must be new, clean and of such a quality as to avoid causing any external or internal damage to the produce.

ii) The use of materials particularly of paper or stamps bearing trade specifications is permitted provided the printing or labeling has been done with non-toxic ink or glue.

iii) The containers shall meet the quality, hygiene, ventilation and resistance characteristics to ensure suitable handling, shipping and preserving of the fruits. Packages must be free of harmful foreign matter and obnoxious smell.

iv) Contents of each package or lot must be uniform and contain only fruits of same origin, variety and grade designation.

v) The visible part of the contents of the package (if present) must be representative of the entire content.

vi) The grade designation mark shall be securely affixed to or printed on each package in a manner approved by the Agricultural Marketing Adviser or an officer authorised by him.

vii) The following particulars shall be clearly and indelibly marked on each package, namely: (a) Name of the commodity, (b) Variety, (c) Grade designation, (d) Size code, (e) Lot/batch/code number, (f) Country of origin, (g) Net weight/No. of fruit, (h) Name and address of the packer/exporter, (i) Best before date (where applicable), (j) Storage condition, (k) Date of packing, Such other particulars as may be specified by the Agricultural Marketing Adviser.

The ink used for marking on packages shall be of such quality which may not contaminate the product. The authorised packer may, after obtaining the prior approval of the Agricultural Marketing Adviser, mark his private trade mark or trade brand on the graded packages provided that the same do not indicate quality other than that indicated by the grade designation mark affixed to the graded packages in accordance with these rules.

3.2 Functions of a Package

Package performs several functions and some of them are described below:

i) As an efficient handling unit, a package is easy to be handled and serves as a marketable unit *e.g.* units with the same content and weight.

ii) Package protects fruit from rough handling during loading, unloading and transport. It protects fruit from pressure during stacking and from moisture or water loss with consequent weight and appearance loss. It also facilitates air flow through crate or box via ventilation holes and fumigation is also possible through ventilation holes.

iii) It facilitates identification(a label with country of origin, volume, type or variety of product, etc. printed on package serves important purpose in marketing).

iv) It provides recognizable trade name and trademark serves purpose of advertisement and help both packer and consumer. It also guarantees quality.

Proper packaging improves appearance of produce. Standard weight and count of a produce in package increases speed and efficiency of marketing. Packaging reduces cost of transport and handling. Stacking and combining of packages into layered units like pallets make mechanical handling possible. It ensures efficient use of space and reduced losses and also lowers the marketing costs. Labels on packages facilitate inspection, pricing and dispute redressal.

3.3 Types of Packaging

Very diverse type of material is used for packaging and it includes paper, plastic, wood, jute and even iron boxes. Trunks made of iron sheets are used for bulk transportation of high value Alphonso mangoes from farms near Ratnagiri (Maharashtra) to wholesale market in Mumbai. Packaging has several advantages and the most important is hygiene.

Fruit needs to be packed for bulk handling, transportation and wholesaling. It reduces losses considerably as compared to transportation of loose/unpacked fruit in the truck. For this purpose, package unit is bigger. For retailing, fruit is packed in smaller unit packages which are also called as consumer packages and the process is called prepackaging.

The packages for fresh fruits are classified as:

i) Transport or Bulk packs

ii) Consumer or Retail packs

3.3.1 Bulk Packaging

Transport packages are designed for long distance transportation in capacities ranging from 4 – 5 kgs to 20 – 25 kgs. These packs must withstand impacts, compression and vibration during transport.

a) Sacks and Net Bags

The materials used for sacks and net bags may be woven natural fibres (jute, cotton), woven synthetic fibres (polypropylene, polyethylene), knitted natural fabric (cotton), knitted synthetic fabric (polyethylene) or non-woven synthetic film (propylene).The advantages of using sacks and net bags are merely financial. The sacks and net bags are cheap, have a low weight/volume ratio and, if made of a synthetic material, do not rot. These sacks usually contain 15-20 kg produce.

b) Wooden Boxes

Wooden Boxes provide better protection to the fresh produce against transportation hazards. They have high puncture resistance, good tensile strength and compression strength; but they occupy more space and add on to the tare weight. Also, the nails cause injuries to the produce during long transportation. A wooden box consists of rigid corners with planks nailed or stretched against those corners. Plank thickness varies normally between 3 and 5 mm.

i) Nailed boxes as generally used for apple or pear are rigid and strong. These are formed just before packing by assembling sides and ends.

ii) Wire bound wooden boxes/crates are stitched with a wire under the stitches which gives extra strength to the container. The wire also serves as a hinge and as a lock for the lid as in case of orange or grapefruit crate.

However, the use of wooden boxes is discouraged due to campaign against deforestation.

c) Fibreboard Boxes

Fibre-board boxes are widely used because of their low weight, their range of sizes and shapes, availability, and versatility (foldable). The other advantages are: a) Low cost to strength and weight ratio, b) Good cushioning properties, c) Smooth and non-abrasive surface, d) Good printability on the outer surface of the board, e) Easy to set up and collapsible for storage, f)Reusability and recyclable, g) Can be manufactured in high volumes and h) Can be provided with ventilation by punching holes.These are of two types *e.g.* Solid fiberboard and Corrugated fiberboard.

Solid fiberboard boxes (cartons) have a thickness between 0.85 and 3 mm. If treated with wax or resin coatings, these boxes are reasonably moisture resistant.

Corrugated fiberboard boxes have a thickness varying from 1.2 up to 8 mm. The strength of corrugated fiberboard is determined by the type of fluting material, the type of facing material and its thickness and a single or double wall. Fluting and facings are kept in place by water resistant glue. These boxes are manufactured in various designs. Most common are one piece Regular Slotted Container (RSC) and two piece full telescopic container (Figure 1).

Figure 1: Telescopic Corrugated Fibre Board Container

Corrugated fiberboard boxes can be arranged on the wooden pallet for handling by forklift trucks for mechanized operations. Boxes can be arranged in a cross stacking or column stacking manner.

d) Plastic Crates

Plastic crates are more expensive than wooden crates or carton boxes, but as a result of their longer life span, the running costs are relatively low. These are usually made of high density polyethylene (HDPE) or polypropylene (PP) by injection moulding. Polyethylene has higher impact strength and a low degradation by ultra-violet radiation while polypropylene has a better scratch resistance. The performance of both materials can be improved by adding antioxidants and UV protectants (for sunlight protection).

The vented plastic crates used for various fruits during handling are of 20-25 kg capacity and hence can be handled manually. Mango, banana, citrus, pomegranate and guavas are handled in field and packinghouses in these crates. These crates are even used for long distance transportation. These are either of stackable, stack-nest or collapsible design. Collapsible plastic crates are the most expensive crates followed by stack-nest and then the stacking crates. The collapsible crates reduce the storage space requirement and transport cost of empty containers. The normal capacity varies between 20 – 40 kgs.

The advantages of plastic crates are: a) Being strong and rigid these can be used for many trips, making the cost per trip relatively low, b) Different sizes and shapes are available, c) They can be easily cleaned and disinfected and d)They are strong and weather resistant and thus they can be used in humid areas

e) Bins

Large size bins or boxes (400-600 kg capacity) are used for more efficient use of available storage spaces compared to smaller crates. Increased speed of mechanical harvesting and handling is another advantage. These bins are handled with forklift trucks for lifting and stacking.

f) Pallet of Corrugated Boxes

These can be mechanized or manual and are unitized for easy mechanical handling with fork-lift trucks. For shipping fresh fruits, the unit size of 120 x 100cm pallet is recommended by organization of economic cooperation and development (OECD). Wood is most preferred material for pallets. Usually pallets designed for 4-way access are preferred. Reusable pallets can be used only if they are made to requirements of EEC standards for use in European Pallet pool system. Non-metal straps for securing and stabilizing the pallet load are used. Corner boards are also used to immobilize the boxes in the pallet (Figure 2). The load is generally 1 metric tonne or little more per pallet.

Figure 2: Wooden Pallet Board at the Bottom with Stacked Corrugated Boxes in Pallet form for Unit Handling

3.3.2 Pre-Packaging or Consumer Packaging

This involves preparation of the fresh produce and packing the same in small marketable units of packages. Fruits like oranges, lemon, banana and grapes can be prepackaged to obtain a one and a half to two times extension in shelf-life in polyethylene bags under normal conditions, without any refrigeration requirements. In the prepackaged condition the product remains fresh without any significant change in nutritional value and taste. The prepackaged produce presents better consumer appeal and has longer shelf-life and considerable handling advantages in transport and marketing.

Prepackaging can be carried out at the farm level or in orchard itself to obtain good results. The major facility required for the prepare at the is a packing shed where these operations can be carried out. Prepackaging in combination with evaporative cooling of the storage space can really benefit the individual farmers or farm cooperatives by extending the marketing period of the fresh produce.

Consumer packages are small in size and designed to hold ½ dozen – 1 dozen fruits weighing about 2-4 kg. Many types of packages in terms of forms and materials are used as consumer packs. The selection criterion for the type of consumer pack depends on fruit characteristic and marketing strategy for the product. The most commonly used packages are listed below:

a) Active Packages
b) Corrugated Boxes/Small cartons
c) Foam Sleeve
d) Modified Atmosphere Packaging (MAP)
e) Paper Bags
f) Plastic Bags made of Flexible Plastic Films
g) Plastic films (Wrapping)
h) Plastic Net/Mesh bags
i) Plastic punnets
j) Smart packages

a) Active Packages

Packaging is termed as "Active" when it performs some desired role other than protecting the contents and providing an inert barrier to the external environment. The goal of developing such packaging is to create a more ideal match of the properties of the package to the requirements of the food. Active Packaging can be created by using oxygen scavangers, carbon dioxide absorbents/emitters, ethanol emittors and ethylene absorbents. The appropriate material is placed alongside the fresh produce. It modifies the atmosphere in the package and thereby contributes to the extension of shelf-life of the fresh fruit. Pads are used to provide chemical treatment to reduce decay as with sulphur dioxide pads for grapes and absorb ethylene as with potassium permanganate pads in boxes of bananas.

b) Corrugated Boxes/Small Cartons

Corrugated fibre board cartons with or without handle and containing 2-4 kg produce are light weight and are used for packaging of apples, grapes, mandarins and mangoes. These are printed in multi-colours with all desired information about produce and packer and are very handy for retail marketing. These can also be opened by the consumers for inspection while buying fruit. Such cartons are quite popular in case of some fruitsas gift packages.

c) Foam Sleeve

It is a plastic tubular film made of polyethylene foam available in different colours, diameters and lengths. It can be easily slipped over the individual fruits in a snug fit form. It provides a cushioning effect and protects the fresh produce against abrasion and scratches during transit. It is hygienic, nontoxic and odourless.

d) Modified Atmosphere Packaging (MAP)

It is a process which involves modifying the composition of the internal atmosphere in air-tight sealed package by use of permeable film to improve the shelf life of produce. The modification process often tries to lower the amount of oxygen

(O_2) from 20 per cent to 8-10 per cent as the respiring fruit consumes O_2. The gaseous atmosphere changes continuously throughout the storage due to factors like respiration and other biochemical changes of the produce and the permeation of gases through the package. This also slows down the growth of aerobic organisms and the speed of oxidation reactions. The CO_2 level inside the package increases by the process of respiration of fruit. CO_2 can also lower the pH and inhibit the growth of bacteria. The film used is permeable to oxygen and carbon dioxide while it is sparsely permeable to water vapour. Therefore, due to higher relative humidity inside the package, produce remains fresh.

MAP reduces respiration, delays ripening, decreases ethylene sensitivity, decreases production of ethylene, retards textural softening and causes extended shelf life of the product. The crucial factor responsible for modified atmospheric storage of fruit is the permeability of the packaging film and respiration rate of the fruit. The permeability of the packaging film is mainly dependent on the thickness of the film and the type of polymer. As the respiration rate of individual fruit varies from each other, the type of polymer film used for MAP also varies. The packaging material slows down the normal respiration of the product by creating a lowered level of O_2 and a higher level of CO_2 inside the packaging material. Improper packaging material may deteriorate the fruit condition and hence reduce its shelf life. For higher shelf life of the product packaging material should be such that optimum atmosphere would be created in the package within the shortest possible time. Various types of flexible and permeable films differ in their CO_2 and O_2 permeability.

The MAP system is a dynamic process where respiration and permeation occur simultaneously. Factors affecting both of these must be considered when designing a package. The fruit respiration in a package is influenced by: 1) Quantity of the produce, 2) Stage of maturity, 3) Temperature, and 4) Concentration of gases like oxygen, carbon dioxide and ethylene. While the permeation of the film is influenced by: 1) Type and nature of material, 2) Thickness and surface area of the material/package, 3) Temperature and relative humidity and 4) Partial pressure gradients of O_2 and CO_2

All these above factors interact to create equilibrium levels of O_2 and CO_2 inside the sealed package. The selection of a film or a combination of more than one film (laminate) depends on the expected transpiration and respiration rates of the produce. For most of the fresh produce the film selected should be more permeable to CO_2 than to O_2.

e) Paper Bags

Consumer packs for 1-2 kg fruits can also be made of good quality kraft paper. These can be suitably printed and provided with jute fibre strings for handle.

f) Plastic Bags made of Flexible Plastic Films

Plastic bags made of different types of flexible plastic films like LDPE (Polyethylene), PVC (Poly Vinyl Chloride), PP (Polypropylene) and cellulose acetate are the predominant material for fruit consumer packaging. Thesebags are mostly used as pouches with holes punched at regular intervals to allow respiration. They are available in a wide range of thicknesses and grades and can be engineered to control the environmental gases inside the pouch. LDPE is the most widely used material.

Besides the very low material costs, automated bagging machines further reduce packing costs. Film bags are clear, allowing for easy inspection of the contents, and readily accept high quality graphics for printing. Plastic films are available in a wide range of thicknesses and grades and may be engineered to control the environmental gases inside the bag.

g) Plastic Films (Wrapping)

One of the newest trends in produce packaging is the shrink wrapping of individual produce items or wrapping with stretchable cling film. The greatest advantage of individual shrink wrapping is its ability to control moisture loss. By reducing the transpiration rate and maintaining the fruit firmness the film forms a barrier which increases the resistance to water vapour. The transpiration rate can be reduced 5 to 20 times using selective permeable plastic films. The individual fruit is loosely sealed in a flexible film. The film is then shrunk tightly around the produce by passing these packs through a heat shrunk tunnel where they are exposed to hot blown air for a very short period (few seconds). The fresh produce is then cooled by rapid ventilation. The films most commonly used are LDPE or LLDPE.

Shrink wrapping has been used successfully to package citrus, apple, pomegranate, mango and a variety of tropical fruits which are large in size.

The overall advantages of individual fruit shrink wrapping are:

i) Reduction in weight loss

ii) Maintenance of firmness

iii) Reduction in deformation

iv) Alleviation of chilling injury

v) Reduction of decay from secondary infection

vi) Delay in colour development and senescence

vii) Elimination of the need to maintain high humidity during storage/transit

viii) Extending the market season

ix) Increased consumer appeal

x) Hygiene

Trays with overwrap of film shrunk with heat using heat shrinkable film are also used. The stretch cling film can also be overwrapped on the fruit. The trays used are usually made of moulded pulp tray or plastic material like EPS, PVC and PP. The produce is placed in individual cavities so that abrasion and bruising are avoided during transportation. The trays also provide cushioning effect to the produce. The overwrap cling film is a transparent see through food-grade, odourless plastic film with the property of clinging to the product packed when stretch wrapped. This film can be applied without the application of heat. It is usually made of LDPE, LLDPE or PVC. The films are semi-permeable and allow exchange of gases for respiration of the product.

h) Plastic Net/Mesh Bags

The plastic net bags (Extruded and Woven) have the feature to stretch and accommodate all sizes and shapes of fruit. These are available in roll form or in precut lengths with stretch width of 200 mm – 400 mm. By allowing air to circulate in and around the produce, these net bags prolong the freshness and shelf-life of the fresh produce. There is no water condensation thereby preventing spoilage and wastage. The bags can be loosely tied at about 5 cm below the opening. These bags make a colourful display by allowing clear visibility of the fruit, enhancing the appeal through natural colours of fresh fruit. These are generally made of High Density Polyethylene (HDPE) or Polyamide (PA).The sturdy mesh bags have much wider use. Citrus and ber fruits are commonly sold in mesh bags. In addition to low cost, mesh bags have advantage of uninhibited air flow and can be kept in cool chamber.

i) Plastic Punnets

These small containers made of transparent PET, PVC or PP (polypropylene) material and holding 250-500 g fruit are used for strawberries, sweet cherries and other small fruits. These are strong, versatile, clear, bright containers, which offer product visibility and are provided with holes for ventilation, which keep the produce fresh. These punnets can be bar coded with price labels and all other desired information. These containers are food-grade, odourless, light weight, stackable and recyclable and give good presentation.

j) Smart Packages

These retail containers are coated with certain patented chemical material that is sensitive to volatiles emanated by spoilage causing bacteria and fungi in fruits. As the fruit starts decaying in package due to growth of spoilage organisms, the container colour changes due to coating which indicates that packed produce is no longer good for consumption.

3.3.3 New Material in Packaging

While plastics can be used because they are easy to manufacture, cheap anddo not rot,the same qualities make these a huge pollution hazard. The problem comes from indiscriminate use and subsequent disposal, particularly the throwaway plastic material used in wrapping and packaging. Efforts are being made to develop eco-friendly sustainable material for packaging of produce including fruits. Verma (2009) reported jute fibre reinforced laminated paper composite and reinforcement-fibre free paper laminate to combat the war against this pollutant to certain extent. The paper laminate, without reinforcement fibre, exhibited a few fold superiority in tensile properties as compared to single paper strip. Further studies showed that an appreciable improvement in tensile properties can be achieved by introducing continuous jute fibre in paper laminates.

Jute fibres and wood are combined with thermoplastic materials like low density polyethylene (LDPE) and high density polyethylene (HDPE) for making semi-rigid as well as rigid boards. The film layer is placed on both sides of non-woven jute. The composite layers are pressed in a hydraulic press at the required temperature for a certain period of time depending upon the thickness of the end product and the type of film used. These boards have good tensile strength, puncture resistance and moisture barrier property. JRP are now being increasingly used for transport packaging of horticultural products.

4.0 Commodity Specific Grading and Packing

Grading and packing standards as mentioned under Fruits and Vegetables Grading and Marking Rules of "Agmark " and amended in 2004, 2007 and 2012 are given below for some important fruits grown in India.

4.1 Aonla (*Emblica officinalis*)

4.1.1 Minimum Requirements

Aonla shall be sufficiently developed and display satisfactory ripeness. It shall have the size, shape and colour characteristics of the variety. Aonla shall be fresh in appearance, whole, with characteristic shape of the variety, sound, clean, free of any visible foreign matter, free of damage or any defect or smell or condensed moisture caused by low temperature or high temperature. The stalk shall be removed close to the fruit. Fruit shall comply with the residue levels of heavy metals, pesticides and other food safety parameters as laid down by the Codex Alimentarius Commission.

4.1.2 Grade Designation

Aonla are graded in three grades:

i) Extra class

ii) Class – I

iii) Class – II

Fruit of "Extra class" must be of superior quality without any type of surface defects. The tolerance limit for this class is 5 per cent by number or weight of fruit not satisfying the requirements of the grade, but meeting those of "Class I" or, exceptionally, coming within the tolerances of that grade.

For "Class I", fruit must be of good quality with surface defects covering 10 per cent area but not affecting pulp. Tolerance is 10 per cent by number or weight of fruit not satisfying the requirements of this grade, but meeting those of next class.

For "Class –II", fruit must meet minimum requirement of quality and surface defects covering 15 per cent area are allowed but it should not be affecting pulp. Tolerance is 10 per cent by number or weight of fruit not satisfying the requirements of this grade but meeting the minimum requirements can be included.

4.1.3 Provision Concerning Sizing

Size is determined by the maximum equatorial diameter of Aonla fruit (Table 1).

Table 1: Aonla Size Code with Fruit Diameter

Size Code	*Maximum Equatorial Diameter in mm.*
A	35.1 and above
B	30.1 – 35.0
C	25.0 – 30.0

4.1.4 Size Tolerance

For all grades, 10 per cent by number or weight of aonla fruit corresponding to the size immediately above and/or below than indicated in the package shall be allowed."

4.1.5 Packaging

i) While packaging the development and condition of Aonla fruit shall be such as to enable them to withstand transport and handling and to arrive in satisfactory condition at the place of destination.

ii) Package should be marked as-fruit for table purpose or for medicinal purpose;

iii) The package of Extra class must contain fruit of uniform of colour and ripeness.

iv) Packages must be free of foreign matter.

v) Aonla may be presented – as arranged in layers, *i.e.*, one or two layers separated from each other or loose in the package (including in bulk bins).

vi) Fruit shall be transported in bulk in the means of transport in bulk packaging, except for the 'Extra Class'.

vii) Aonla in 'Extra Class' shall be particularly well packed and presented.

4.2 Apple (*Malus domestica*)

4.2.1 Minimum Requirements

(a) Apples shall be:

i) Whole, the stalk should be intact up to the shoulder height, the cut should be clean and should not damage the adjacent skin;

ii) Firm;

iii) Sound, not affected by rotting or deterioration which makes it unfit for consumption;

iv) Clean, free from any visible foreign matter;

v) Free from scabs;

vi) Free from pests and damage caused by them or diseases affecting the general appearance of the produce;

vii) Free from abnormal external moisture such as condensation following removal from cold storage;

viii) Free from any foreign smell or taste, damage caused by low or high temperature, skin puncture, signs of dehydration and internal blackening or rotting.

(b) These have minimum total soluble solids of 11 degree brix and apples shall comply with the residue levels of heavy metals, pesticides and other food safety requirements as laid down by the Codex Alimentarius Commission or requirement of importing country.

4.2.2 Grade Designation

Apples are graded in three classes as per their grade (Table 2).

Table 2: Grade Designation for Apples with Grades

Grade Designation	*Grade Requirements*	*Grade Tolerance*
Extra class	Apples in this class must be of superior quality. The flesh must be sound. They shall be characteristic of the variety and they must be free of defects, with the exception of very slight superficial defects, provided these do not affect the general appearance of the produce, the quality, the keeping quality and presentation in the package. (five percent of the fruits may have the defects as given in table for tolerance of defects)	Five percent by number or weight of apples not satisfying the requirements of the grade, but meeting those of Class-I or, exceptionally, coming within the tolerances of that grade. Included therein shall be allowed not more than 1% for apples affected by internal breakdown.
Class – I	Apples in this class must be of good quality. The flesh must be sound and they must be characteristic of the variety. Provided that the following do not amount to defect if they do not affect the general appearance of the produce, the quality, the keeping quality and presentation in the package which have: ☆ A slight defect in shape and development; ☆ A slight defect in colouring; ☆ Slight skin or other defects. (ten percent of the fruits may have the defects as given in table for tolerance of defects)	Ten percent by number or weight of apples not satisfying the requirements of the grade, but meeting the requirements of grade II. Included therein shall be allowed not more than 1% for apples affected by decay or internal breakdown
Class – II	This class includes apples, which do not qualify for inclusion in the higher classes, but satisfy the minimum requirements. Provided that the following do not amount to defect if the apples retain their essential characteristics as regards the quality, the keeping quality and presentation which have: ☆ Defects in shape and development; ☆ Defects in colouring; ☆ Skin or other defects. (ten percent of the fruits may have the defects as given in table for tolerance of defects).	Ten percent by number or weight of apples not satisfying the requirements of the grade, but meeting the minimum requirements, with the exception of apples affected by decay or internal breakdown that should not be more than two percent. Included therein shall be allowed, maximum of 2% by number or weight of fruits which may show the following defects: ☆ Cork like blemishes(bitter pit) ☆ Slight damage or unhealed broken skin/cracks.

Note: In each class apples must have been carefully picked and have colour that is characteristic of the variety.

In each grade/class, certain tolerance limits have been fixed to accommodate fruit with slight superficial blemishes (Table 3).

Table 3: Tolerance for Defects

Defects Allowed	*Extra Class*	*Class I*	*Class II*
Russetting outside Calyx/stem cavity, net like	1 per cent of surface area	3 per cent of surface area	5 per cent of surface area
Smooth solid russeting	1 per centof surface area	1 per cent of surface area	3 per cent of surface area
Accumulation for both types of russetting should not exceed	1 per cent of surface area	3 per cent of surface area	5 per cent of surface area
Accumulated Blemishes and Bruising including-Within this limit scar caused by scabs* (*Venturia inaequalis*) shall not exceed.	Nil	0.25 sq.cm area	0.5 sq.cm area
Stem or Calyx cracks (healed or well cured)	Nil	0.5 cm	1 cm
Maximum length of elongated shaped defects	Nil	1 cm	2 cm

Note:

☆ "blemish" means superficial bruising which is not covered under other defects.

☆ "flesh defects" means visible water core and internal or external bitter pit.

☆ "other skin defects" means defects due to hail (healed), limb damage, dark brown or black marks produced by branches, insect pest and disease spot not resulting in skin punctures.

☆ "russetting" means particular type of skin, abnormal roughening or streak and scarring of the fruit surface, which characteristically is smooth.

4.2.3 Provisions Concerning Sizing

Size code is determined by the diameter in millimeter of the equatorial section or weight of the fruit in grams in accordance with the Table 4.

Table 4: Apple Size Code, Fruit Diameter and Fruit Weight

Size Code	*Fruit Diameter (mm)*	*Fruit Weight (g)*
A	More than 80	More than 241
B	76-80	211-240
C	71-75	181-210
D	66-70	151-180
E	61-65	121- 150
F	56-60	91-120
G	50-55	60-90

4.2.4 Size Tolerance

For all classes, 10 per cent by number or weight of apples not satisfying requirements of the size code.

Coloured apple varieties shall fulfill colour criteria to be included in to the given Class (Table 5). Following shall be the colour criteria for different grades of apples of different varieties except for green and yellow apple varieties.

Table 5: Colour Criteria for Apples

Grade	*Percentage characteristic colour of the variety*
Extra Class	85 or more
Class – I	70 or more
Class – II	Less than 70

4.2.5 Packaging

i) The contents of each package must be uniform and contain only apples of the same origin, variety, quality, size and same degree of ripeness.

ii) The Extra Class fruits must be packed in layers only.

iii) The apples in Class-I and Class-II may be presented in one of the following forms, namely:

- In regular layers according to size;
- In open or closed containers;
- Loose in consumer packages;
- Shipping packages (master containers or cartons) or
- In bulk or pallet bins.

iv) Consumer packages (of net weight not exceeding 5 Kg) may contain mixture of varieties, provided they are of uniform grade designation and each variety is of the same origin. The visible part of the content of the package must be representative of the entire content.

v) Other than bulk or pellet packing, the maximum net weight of the fruits shall not exceed 20Kg.

vi) Stickers shall be affixed to each produce so that, when removed, they neither leave visible traces of glue, nor lead to skin defects.

Himachal Pradesh Horticultural Produce Marketing and Processing Corporation Ltd. popularly known as" hpmc" recommended two Quality Grades of apple *i.e.* Extra Fancy and Fancy Class -1 are to be offered for export. Himachal Pradesh produces premium quality apples of varieties- namely Royal Delicious, Red Delicious, Rich-A-Red Delicious and Golden Delicious. The quality of Himachal Apple meets the entire export standards.

The broad parameters of the two Quality grades for export are that the fruit must be sound, fully developed, clean, without any surface blemishes, coloured, juicy and full of flavour and freshness.

Tray Pack Cartons have specially been developed for export of Himachal Apples (Figure 3). It consists of an inner fibre board carton, a series of moulded trays and an outer telescopic lid which fits right down over the sides and end of the inner carton. Apples in Tray Pack cartons are wrapped as per requirement of the buyer. The net weight of apples in the carton is not less than 17 kg but in practice it is usually over 18 kg depending on the variety. Up to 80 apples of extra large size and 175 apples of extra small size can be packed in these cartons (Table 6). The inner dimensions of the carton are 500x300x278 (ht) mm and inside dimensions of the outer cover are 504x303x281.mm.Apples are also packed in a single layer in open top carton with each fruit placed in polystyrene foam sleeve to avoid rubbing (Figure 4).

Figure 3: Apples Placed on Moulded Trays and Packed in Carton

Figure 4: Apples Packed in Polystyrene Foam Sleeve and Packed in Open Top Carton

Table 6: Apple Size, Fruit per Tray, Number of Trays per Box and Fruit per Carton

Size of Apple	*No. of Apple Pocket in a Tray*	*No. of Trays in a Carton*	*No. of Apples in a Carton*
Extra Large (80 to 85 mm)	20	4	80
Large (75 to 80mm)	20	5	100
Medium (70 to 75mm)	25	5	125
Small (65 to 70mm)	30	5	150
Extra Small (60 to 65mm)	35	5	175

4.3 Banana (*Musa paradisica*)

4.3.1 Minimum Requirements

In all classes, subject to the special provisions for each class and the tolerances allowed, the bananas must be:

i) Whole (taking the finger as the reference);

ii) Sound, produce affected by rotting or deterioration such as to make it unfit for consumption is excluded;

iii) Clean, practically free of any visible foreign matter;

iv) Practically free of pests affecting the general appearance of the produce;

v) Practically free of damage caused by pests;

vi) Free of abnormal external moisture, excluding condensation following removal from cold storage, and bananas packed under modified atmosphere conditions;

vii) Free of any foreign smell and/or taste;

viii) Firm;

ix) Free of damage caused by low temperatures, bruises, malformation or abnormal curvature of the fingers, with pistils removed, with the stalk intact, without bending, fungal damage or desiccation.

In addition, hands and clusters must include:

i) A sufficient portion of the crown of normal colouring, sound and free of fungal contamination;

ii) A cleanly cut crown, not bevelled or torn, with no stalk fragments.

Bananas shall comply with the residue levels of heavy metals, Pesticides and other food safety parameters as laid down by the Codex Alimentarius Commission for exports.

4.3.2 Grade Designation

Bananas are graded in three classes as per their quality (Table 7).

4.3.3 Provisions Concerning Sizing

For the purposes of sizing bananas, the length of the fingers is determined along the outside curve from the blossom end to the base of the pedicel where the edible pulp ends and the diameter is defined as the thickness of a transverse section between the lateral faces. The reference fruit for measurement of the length and grade is:

i) For hands, the median finger on the outer row of the hand,

ii) For clusters, the finger next to the cut section of the hand, on the outer row of the cluster.

The minimum length of finger should not be less than 14.0 cm and the minimum grade not less than 2.7 cm.

4.3.4 Size Tolerance

For all classes, 10 per cent by number or weight of bananas not satisfying the sizing characteristics, up to a limit of 1 cm for the minimum length of 14 cm.

For export, bananas are recommended to be packed in 13 kg carton (NHB, 2011) and fruit should be green (Table 8).

Table 7: Criteria for Grade Designation of Banana

Grade Designation	*Grade Requirements*	*Grade Tolerance*
Extra class	Bananas shall be of superior quality. They must be characteristics of the variety and/or commercial type. The fingers must be free of defects, with the exception of very slight superficial defects, provided these do not affect the general appearance of the produce, quality, the keeping quality and presentation in the package.	5 per cent by number or weight of bananas not satisfying the requirements of the grade, but meeting those of for Class I grade or, exceptionally, coming within the tolerances for that class.
Class – I	Bananas shall be of good quality. They must be characteristics of the variety and/or commercial type. The following slight defects of the fingers, however, may be allowed, provided these do not affect the general appearance of the produce, quality, the keeping quality and presentation in the package. ☆ Slight defects in shape and colour; ☆ Slight defects due to rubbing and other superficial defects not exceeding 2 sq.cm. of the total surface area. The defects must not affect the flesh of the fruit.	10 per cent number or weight of bananas not satisfying the requirements of the grade class, but meeting those of Class II or, exceptionally, coming within the tolerances of that grade.
Class – II	This includes bananas which do not qualify for inclusion in the higher classes, but satisfy the minimum requirements. The following defects may be there, provided the bananas retain their essential characteristics as regards the quality, the keeping quality and presentation. ☆ Defects in shape and colour provided the product remains the normal characteristics of bananas; ☆ Skin defects due to scrapping, scabs, rubbing, blemishes or other causes not exceeding 4 sq.cm. of the total surface area; The defects must not affect the flesh of the fruit.	10 per cent by number or weight of bananas not satisfying the requirements of the grade, but meeting the minimum requirements.

Table 8: Specification for Packaging of Banana for Export in Middle East

Variety	Grand Naine, Cavendish
Colour	Green
Weight of hand	2.5 kg
Packing	13 kg

Storage	13-14 °C
Transport	By Sea

4.3.5 Packaging

i) While packaging development and condition of the bananas must be such as to enable them to reach the appropriate stage of physiological maturity corresponding to the particular variety to withstand transport and handling, and arrive in satisfactory condition at the place of destination in order to ripen satisfactorily.

ii) Bananas must be packed in hands and clusters (parts of hands) of at least four fingers. Bananas may also be presented as single finger.

iii) Clusters with no more than two missing fingers are allowed, provided the stalk is not torn but clearly cut, without damage to the neighboring fingers.

iv) Not more than one cluster of three fingers with the same characteristics as the other fruit in the package may be present per row.

4.4 Cherry (*Prunus avium*)

4.4.1 Minimum Requirements

Cherries should have stem attached.

Missing stems are allowed in the case of :

i) **"Sweet varieties"** which naturally lose the stem at harvest, on condition that the skin is not damaged.

ii) **"Sour varieties"** on condition that there is no severe leakage of juice.

Cherries should comply with the residue levels of heavy metals, pesticides and other food safety parameters as laid down by the Codex Alimentarius Commission for exports.

4.4.2 Grade Designation

Cherries are graded in three grades:

i) Extra class

ii) Class – I

iii) Class – II

Fruit of Extra class must be of superior quality.

4.4.3 Provisions Concerning Sizing

Size is determined by the maximum diameter of the equatorial section of the fruit (in millimeters)

Cherries must have following minimum sizes: 20 mm for Extra Class and 17 mm for Class I and II.

4.4.4 Size Tolerance

For all grades, 10 per cent by number or weight of cherries not conforming to the minimum size specifies, provided, that the diameter is not less than

i) 18 mm. in the "Extra class"

ii) 15 mm. in class I and Class II.

4.4.5 Packaging

i) At the time of packaging, cherries must have been carefully picked and have reached an appropriate degree of development and displayed satisfactory ripeness in accordance with criteria proper to the variety and to the area in which they are grown.

ii) While packing, development and condition of the cherries must be such as to enable them to withstand transport and handling, and to arrive in satisfactory condition at the place of destination.

iii) Cherries marked as 'Extra' class on the package must be of uniform colouring and ripeness.

iv) The packages must be free from foreign matter.

v) For sour cherries, the package may be marked 'sour cherries'.

4.5 Custard Apple (*Annona squamosa*)

4.5.1 Minimum Requirements

At the time of maturity, custard apple colour between the areoles should be creamy white.

Fruit shall comply with the residue levels of heavy metals, pesticides and other food safety parameters as laid down by the Codex Alimentarius Commission for exports.

4.5.2 Grade Designation

Custard apple are graded in three grades:

i) Extra class

ii) Class – I

iii) Class – II

Fruit of "Extra class" must be of superior quality without surface defects. The tolerance limit for this class is 5 per cent by number or weight of fruit not satisfying the requirements of the grade, but meeting those of "Class-I" or, exceptionally, coming within the tolerances of that grade.

For "Class I", fruit must be of good quality with surface defects covering 10 per cent area but not affecting pulp. Grade tolerance is 10 per cent by number or weight of fruit not satisfying the requirements of the grade, but meeting those of next class. For "Class –II", fruit must meet minimum requirement of quality and surface defects covering 15 per cent area are allowed but not affecting pulp. Grade tolerance is 10 per cent by number or weight of fruit not satisfying the requirements of the grade but meeting the minimum requirements can be included.

4.5.3 Provision Concerning Sizing

Size is determined by the weight of the fruit (Table 9).

Table 9: Custard Apple Size Code and Fruit Weight

Size Code	*Weight (g)*
A	150-200
B	201-300
C	301 and above

Minimum weight of fruits for "Extra class" shall not be less than 200 gms.

4.5.4 Size Tolerance

For all classes, 10 per cent by number or weight of Custard apples not conforming to the minimum size specifies, for the grade, but meeting the size requirement for that grade immediately below.

4.5.5 Packaging

i) Fruit should have been carefully picked and have reached an appropriate degree of development in accordance with criteria proper to the variety and to the area in which they are grown.

ii) The development and condition of the custard apples must be such as to enable to withstand transport and handling, and to arrive in satisfactory condition at the place of destination.

iii) In case of "Extra class" and "Class I" packages, extra care is necessary to avoid any disintegration into segments.

4.6 Grape (*Vitis vinifera*)

4.6.1 Minimum Requirements

i) Bunches and berries of Table grapes shall be: (a) clean, sound, free of any visible foreign matter; (b) free of pests, affecting the general appearance of the produce; (c) free of damage caused by pests or diseases; (d) free of abnormal external moisture; (e) free of any foreign smell and/or taste; (f) free of all visible traces of moulds;

ii) Berries shall be intact, well formed and normally developed;

iii) Table grapes shall comply with the residue levels of heavy metals, pesticides and other food safety parameters as laid down by the Codex Alimentarius Commission for Exports.

iv) Table grapes shall have minimum soluble solids of 16 degrees Brix.

v) Table grapes shall have minimum sugar/acid ratio of 20: 1.

vi) Pigmentation due to sun is not a defect.

4.6.2 Grade Designation

Grapes are classified in three grades depending on their quality (Table 10)

Table 10: Criteria for Grade Designation

Grade Designation	*Grade Requirements*	*Grade Tolerances*
Extra class	Grapes must be of superior quality. The bunches must be typical of variety in shape, development and colouring and have no defects. Berries must be firm, firmly attached to the stalk, evenly spaced along the stalk and have their bloom virtually intact.	5 per cent by weight of bunches not satisfying the requirements for the grade, but meeting those of Class I grade or exceptionally coming within the tolerances of that grade.
Class – I	Grapes must be of good quality. The bunches must be typical of variety in shape, development and colouring. Berries must be firm, firmly attached to the stalk and, as far as possible, have their gloom intact. The may, however, be less evenly spaced along the stalk than in the extra class. Following slight defects may be there, provided these do not affect the general appearance of the produce and keeping quality of the package. ☆ A slight defect in shape, ☆ A slight defect in colouring	10 per cent by weight of bunches not satisfying the requirements for the grade, but meeting those of Class II grade or exceptionally coming within the tolerances of that grade..
Class – II	The bunches may show defects in shape, development and colouring provided these do not impair the essential characteristics of the variety. The berries must be sufficiently firm and sufficiently attached. They may be less evenly spaced along the stalk than Class I grade. Following defects may be there, provided these do not affect the general appearance of the produce and keeping quality of the package. ☆ Defects in shape, ☆ Defects in colouring, ☆ Slight sun scorch affecting the skin only, ☆ Slight bruising, ☆ Slight skin defects.	10 per cent by weight of bunches and satisfying the requirements of the grade, but meeting the minimum requirements..

4.6.3 Provisions Concerning Sizing

Size is determined by the weight of bunches (Table 11). The following minimum (in gms.) requirements per bunch are laid down for large and small berries grapes.

Table 11: Grape-Grades and Minimum Bunch Weight

Grade	*Bunch Weight (g)*	
	Varieties with Large Berries	*Varieties with Small Berries*
Extra class	200	150
Class I	150	100
Class II	100	75

4.6.4 Size Tolerance

Extra class, Class I, Class II : 10 per cent by weight of bunches not satisfying the size requirements for the grade, but meeting the size requirements for the grade immediately below.

After harvesting, preliminary sorting is necessary to remove culled damaged bunches. Bunches of specified weight are packed in CFB cartons of 1, 4.5 or 9 kg capacity (NHB, 2011) depending on the country to which fruit is exported (Table 12).

4.6.5 Packaging

While packaging, grapes must have been carefully picked and have reached an appropriate degree of development and ripeness in accordance with criteria proper to the variety and/or commercial type and to the area in which they are grown. The Development and condition of the Grapes must be such as to enable them to withstand transport and handling, and to arrive in satisfactory condition at the place of destination.

Different types of containers are used for packing grapes in India. These include: i) bucket shaped baskets (36cm. diameter at the top, 20 cm. at base and 25cm. height) of 5 kg capacity and made of thin bamboo strips are commonly used in Andhra Pradesh and Karnataka. ii) Wooden boxes rectangular in shape and capacity to accommodate 5 to 8kg. of fruits are also used. iii) Ventilated corrugated fibre board (CFB) cartons.

Grapes are packed with paper clippings as cushioning which absorb moisture released by the grapes and provide insulation against minor fluctuations in temperature during cold storage. Size of the box varies to accommodate 2 to 4 kg. of fruit. The inner dimensions of the CFB carton accommodating 4 kg. grapes are 37 cm. in length, 25.5 cm in width and 11.5 cm. in height, and those of the carton accommodating 2kg. grapes are 25.5x18.5x11.5 cm. Irrespective of the size of the carton, each one contains two holes of 2cm. diameter on either side.

Table grapes for overseas markets are packed in five-ply corrugated fibre board boxes, printed and labeled attractively. The standard dimensions of the carton are 50 cm. x 30 cm. x 12 cm. to accommodate 5 kg. grapes.

Table 12: Specifications for Grape Export to Europe and Middle-East

Variety	*Colour*	*Berry Size (mm)*			*Packing (kg)*		
		Holland/Germany	*U.K.*	*Middle East*	*Holland/Germany*	*U.K.*	*Middle East*
Thompson Seedless	White/Amber	16	18	15	4.5 or 9	4.5 or 9	1
Sharad Seedless	Black	16	18	15	4.5 or 9	4.5 or 9	1
Flame Seedless	Pink	16	18		4.5 or 9	4.5 or 9	1

Cartons made from recyclable materials, without wax and printed with water based inks in custom design and graphics are now available. Cartons with die cut ends require less energy and save on paper waste also.

Table grapes meant for local market are picked and packed directly in containers in the field. Table grapes meant for long distance markets and those for overseas markets are packed differently in the packing shed.

Grapes are also packed in plastic punnets with a net weight of 500 or 400 g. Trimmed and sorted fruit is punnet packed thereafter punnets are placed into 5 kg cartons. Pallet units contains 115 or 125 cartons.

4.7 Guava (*Psidium guajava*)

4.7.1 Minimum Requirements

The guavas must have been carefully picked and have reached an appropriate degree of development and ripeness in accordance with criteria proper to the variety and to the area in which they are grown. The development and condition of the guavas must be such as to enable them to withstand transport and handling, and to arrive in satisfactory condition at the place of destination.

4.7.2 Grade Designation

Guavas are graded in three grades:

i) Extra class

ii) Class – I

iii) Class – II

"Extra class" fruit must be of superior quality without surface defects. The tolerance limit for this class is 5 per cent by number or weight of fruit not satisfying the requirements of the class/grade, but meeting those of "Class I" or, exceptionally, coming within the tolerances of that grade.

For "Class I", fruit must be of good quality with surface defects covering 10 per cent area but not affecting pulp. Tolerance is 10 per cent by number or weight of fruit not satisfying the requirements of the class/grade, but meeting those of next class.

"Class –II" fruit must meet minimum requirement of quality and surface defects covering 15 per cent of the fruit area are allowed but it should not affect pulp. Ten percent by number or weight of guavas satisfying neither the requirements of the class nor the minimum requirements, with the exception of produce affected by rotting or any other deterioration rendering it unfit for consumption.

4.7.3 Provisions Concerning Sizing

Size is determined by the weight or maximum diameter of the equatorial section of the fruit, in accordance with the Table 13.

Table 13: Guava Fruit Size Code with Corresponding Weight and Diameter

Size Code	Weight (g)	Diameter (mm)
A	> 350	> 95
B	251-350	86-95
C	201-250	76-85
D	151-200	66-75
E	101-150	54-65
F	61-100	43-53

4.7.4 Size Tolerance

For all grades, 10 per cent by number or weight of guavas corresponding to the size immediately above and/or below that indicated on the package.

4.7.5 Packaging

For export from India, guavas are being packed in 5 kg capacity open top carton. There are about 170 cartons in one pallet and gross weight of each pallet is about 850 kg. Twenty such pallets are accommodated in 40 feet long refrigerated container (Reefer) with weight of 17 tonnes.

4.8 Grapefruit (*Citrus paradisi*)

4.8.1 Minimum Requirements

Grapefruit shall be fresh in appearance, whole, with characteristic shape of the variety, sound, clean, free of any visible foreign matter, free of damage or defect or smell or condensed moisture caused by low temperature or high temperature. The stalk shall be removed close to the fruit keeping the button intact. Fruit shall comply with the residue levels of heavy metals, Pesticides and other food safety parameters as laid down by the Codex Alimentarius Commission for exports. Grapefruits shall have minimum diameter of 70 mm.

4.8.2 Grade Designation

Grapefruit are graded in three grades:

i) Extra class

ii) Class – I

iii) Class – II

Fruit of "Extra class" must be of superior quality without surface defects. The tolerance limit for this class is 5 per cent by number or weight of fruit not satisfying the requirements of the grade, but meeting those of "Class I" or, exceptionally, coming within the tolerances of that grade. For "Class I", grapefruit must be of good quality with surface defects covering 10 per cent area but not affecting pulp. Tolerance is 10 per cent by number or weight of fruit not satisfying the requirements of the grade, but meeting those of next class. For "Class –II", fruit must meet minimum requirement of quality and surface defects covering 15 per cent area are allowed but not affecting pulp. Tolerance is 10 per cent by number or weight of fruit not satisfying the requirements of the grade but meeting the minimum requirements can be included.

4.8.3 Provision Concerning Sizing

Size is determined by the maximum diameter of the equatorial section of the fruit in accordance with the Table 14.

Table 14: Grapefruit Size and Diameter

Size Code	*Diameter (mm)*
A	71-80
B	81-90
C	91-100
D	101-110
E	111-120
F	121-130
G	131 and above

4.8.4 Size Tolerance

i) For all classes and types of presentation: 10 per cent by number or weight of Grapefruits corresponding to the size immediately below and/or above that mentioned on the packages provided the maximum difference between the small and largest fruit does not exceed the following maxima:

Maximum difference between fruit in A,B,C and D size code should not be 7 mm and for Size code E,F and G, it should not be more than 10mm.

ii) In case of bulk consignments, 10 per cent only applies to Grapefruits with a diameter of not less than 70 mm.

4.8.5 Packaging

i) At the time of packaging grapefruits shall have been carefully picked and have reached an appropriate degree of development and display satisfactory ripeness in accordance with criteria proper to the variety and to the area in which they are grown.

ii) The development and condition of the Grape fruits shall be such as to enable them to withstand transport and handling, and to arrive in satisfactory condition at the place of destination.

iii) The content of each package must be uniform and contain only grapefruits of the same origin, variety, quality, size code, same degree of ripeness and development.

iv) During packaging, use of any substance tending to modify the natural characteristics of the grapefruits, especially its taste or smell is prohibited.

v) Fruits for "Extra Class" shall be arranged in regular layers in packages. This type of fruit arrangement is optional for Class I and II.

vi) If required, fruits of Classes I and II can be jumble packed.

4.9 Lime (*Citrus aurantifolia*)

4.9.1 Minimum Requirements

Limes shall be fresh in appearance, whole, with characteristic shape of the variety, sound, clean, free of any visible foreign matter, free of damage or defect or smell or condensed moisture caused by low temperature or high temperature. The stalk shall be removed close to the fruit keeping the button intact. Fruit shall comply with the residue levels of heavy metals, Pesticides and other food safety parameters as laid down by the Codex Alimentarius Commission for exports. Limes shall have a minimum diameter of 33 mm.

4.9.2 Grade Designation

Limes are graded in three grades:

i) Extra class

ii) Class – I

iii) Class – II

Fruit of "Extra class" must be of superior quality without surface defects. The tolerance limit for this class is 5 per cent by number or weight of fruit not satisfying the requirements of the grade, but meeting those of "Class I" or, exceptionally, coming within the tolerances of that grade.

For "Class I", limes must be of good quality with surface defects covering 10 per cent area but not affecting pulp. Tolerance is 10 per cent by number or weight of fruit not satisfying the requirements of the grade, but meeting those of next class. For "Class –II", fruit must meet minimum requirement of quality and surface defects covering 15 per cent area are allowed but not affecting pulp. Tolerance is 10 per cent by number or weight of fruit not satisfying the requirements of the grade but meeting the minimum requirements can be included.

4.9.3 Provision Concerning Sizing

Size is determined by the maximum diameter of the equatorial section of the fruit in accordance with the Table 15.

Table 15: Acid Lime Fruit Size and Diameter

Size Code	*Diameter (mm)*
A	33-37
B	38-42
C	43-47
D	48-52
E	53 and above

4.9.4 Size Tolerance

For all classes, 10 per cent by number or weight of limes corresponding to the size immediately below and/or above that mentioned on the packages. In any case, the tolerance of 10 per cent applies only to fruit not smaller than the minimum of 33 mm.

4.9.5 Packaging

i) While packaging, limes shall have been carefully picked and have reached at an appropriate degree of development and display satisfactory ripeness in accordance with criteria proper to the variety and to the area in which they are grown.

ii) The development and condition of the Limes must be such as to enable them to withstand transport and handling, and to arrive in satisfactory condition at the place of destination.

iii) The content of each package shall be uniform and contain only limes of the same origin, variety, quality, size code, same degree of ripeness and development.

iv) The use of any substance tending to modify the natural characteristics of the limes, especially its taste or smell is prohibited.

v) The limes shall be arranged in regular layers in packages; it is mandatory for "Extra Class" and optional for Class I and II.

vi) For Class I and II, it is allowed to pack the fruit without regular layers in packages or in bulk bins. This is called as jumble pack.

4.10 Lemon (*Citrus limon*)

4.10.1 Minimum Requirements

Lemons shall be fresh in appearance, whole, with characteristic shape of the variety, sound, clean, free of any visible foreign matter, free of damage or defect or smell or condensed moisture caused by low temperature or high temperature. The stalk shall be removed close to the fruit keeping the button intact. Fruit shall comply with the residue levels of heavy metals, Pesticides and other food safety parameters as laid down by the Codex Alimentarius Commission for exports. Lemons shall have minimum diameter of 45 mm.

4.10.2 Grade Designation

Lemons are graded in three grades:

i) Extra class

ii) Class – I

iii) Class – II

Fruit of "Extra class" must be of superior quality without surface defects. The tolerance limit for this class is 5 per cent by number or weight of fruit not satisfying the requirements of the grade, but meeting those of "Class I" or, exceptionally, coming within the tolerances of that grade. For "Class I", lemons must be of good quality with surface defects covering 10 per cent area but not affecting pulp. Tolerance is 10 per cent by number or weight of fruit not satisfying the requirements of the grade, but meeting those of next class. For "Class –II", fruit must meet minimum requirement of quality and surface defects covering 15 per cent area are allowed but not affecting pulp. Tolerance is 10 per cent by number or weight of fruit not satisfying the requirements of the grade but meeting the minimum requirements can be included.

4.10.3 Provision Concerning Sizing

Size is determined by the maximum diameter of the equatorial section of the fruit in accordance with the Table 16.

Table 16: Lemon Fruit Size Code and Diameter

Size Code	*Diameter (mm)*
A	45-50
B	51-56
C	57-62
D	63-68
E	75 and above

4.10.4 Size Tolerance

For all classes, 10 per cent by number or weight of lemons corresponding to the size immediately below and/or above that mentioned on the packages. In any case, the tolerance of 10 per cent applies only to fruit not smaller than the minimum of 45 mm.

4.10.5 Packaging

i) At the of packing, lemons shall have been carefully picked and have reached an appropriate degree of development and display satisfactory ripeness in accordance with criteria proper to the variety and to the area in which they are grown.

ii) Fruit development and condition must be such as to enable them to withstand transport and handling, and to arrive in satisfactory condition at the place of destination.

iii) The content of each package shall be uniform and contain only lemons of the same origin, variety, quality, size code, same degree of ripeness and development.

iv) The use of any substance tending to modify the natural characteristics of the lemons, especially its taste or smell is prohibited.

v) The lemons shall be arranged in regular layers in packages; it is mandatory for "Extra Class" and optional for Class I and II.

vi) For Class I and II, it is allowed to pack the fruit without regular layers in packages or in bulk bins. This is called as jumble pack.

4.11 Litchi (*Litchi chinensis*)

4.11.1 Minimum Requirements

Litchis shall be carefully picked and must be sufficiently developed and matured. The development and condition of the Litchis must be such as to enable them to withstand transport and handling and to arrive in satisfactory condition at the place of destination.

4.11.2 Grade Designation

Litchis are graded in three grades:

i) Extra class

ii) Class – I

iii) Class – II

Fruit of Extra class must be of superior quality without surface defects. The tolerance limit for this class is 5 per cent by number or weight of fruit not satisfying the requirements of the grade, but meeting those of Class I or, exceptionally, coming within the tolerances of that grade.

For Class I, fruit must be of good quality with surface defects covering 10 per cent area but not affecting pulp. Grade tolerance is 10 per cent by number or weight of fruit not satisfying the requirements of the grade, but meeting those of next class. For "Class –II", fruit must meet minimum requirement of quality and surface defects covering 15 per cent area are allowed but not affecting pulp. Tolerance is 10 per cent by number or weight of fruit not satisfying the requirements of the grade but meeting the minimum requirements can be included.

4.11.3 Provisions Concerning Sizing

Fruits are graded depending on minimum diameter (Table 17).

Table 17: Litchi Fruit Size and Diameter

Grade	*Minimum Equatorial Diameter (mm) Minimum*
Extra class	33
Class – I	28
Class – II	23

4.11.4 Size Tolerance

i) A maximum size range of 10 mm. between fruits in each package is permitted.

ii) For all grades, 10 per cent by number or weight of litchis not satisfying the requirements as regards the minimum size of grade are allowed, provided, that the diameter is not less than 23 mm.

4.11.5 Packaging

Packed litchis shall be presented under one of the following forms:

i) Individually - In this case the pedicel must be cut at the first knot and the maximum length of the stalk must not extend more than 2 mm. beyond the top of the fruit. Litchis in special grade must be presented individually.

ii) In bunches — In this case the bunch must include more than three attached and well formed Litchis. The branch must not exceed 15 cm. in length.

Fruits are packed in printed corrugated printed fibre - board boxes of required strength. Boxes should be able to withstand high humidity in cold storage during stacking.

4.12 Mandarin (*Citrus reticulata*)

4.12.1 Minimum Requirements

Mandarins shall be fresh in appearance, whole, with characteristic shape of the variety, sound, clean, free of any visible foreign matter, free of damage or defect or smell or condensed moisture caused by low temperature or high temperature. The stalk shall be removed close to the fruit keeping the button intact. Fruit shall comply with the residue levels of heavy metals, Pesticides and other food safety parameters as laid down by the Codex Alimentarius Commission for exports. Mandarins shall have minimum diameter of 54 mm.

4.12.2 Grade Designation

Mandarins are graded in three grades:

i) Extra class

ii) Class – I

iii) Class – II

Fruit of "Extra class" must be of superior quality without any type of surface defects. The tolerance limit for this class is 5 per cent by number or weight of fruit not satisfying the requirements of the grade, but meeting those of "Class I" or, exceptionally, coming within the tolerances of that grade.

For "Class I", mandarins must be of good quality with surface defects covering 10 per cent area but not affecting pulp. Tolerance is 10 per cent by number or weight of fruit not satisfying the requirements of the grade, but meeting those of next class. For "Class –II", fruit must meet minimum requirement of quality and surface defects covering 15 per cent area are allowed but not affecting pulp. Tolerance is 10 per cent by number or weight of fruit not satisfying the requirements of the grade but meeting the minimum requirements can be included.

4.12.3 Provision Concerning Sizing

Size is determined by the maximum diameter of the equatorial section of the fruit in accordance with the Table 18.

Table 18: Mandarin Fruit Size and Diameter

Size Code	*Diameter (mm)*
A	54-61
B	62-69
C	70-77
D	78-85
E	Above 85

4.12.4 Size Tolerance

For all classes and types of presentation: 10 per cent by number or weight of Mandarins corresponding to the size immediately below and/or above that mentioned on the packages, provided the maximum difference between the smallest and largest fruits does not exceed the following maxima.

Maximum difference allowed between fruit in the same package is 7 mm for size Code of A,B,C and D and difference of 10 mm is allowed for size code E. In any case, the tolerance of 10 per cent applies only to fruit whose diameter is not less than 54 mm.

4.12.5 Packaging

The content of each package shall be uniform and contain only Mandarins of the same origin, variety, quality, size, same degree of ripeness and development.

The use of any substance tending to modify the natural characteristics of the Mandarins, especially its taste or smell is prohibited.

The Mandarins shall be presented:

i) Arranged in regular layers in packages. It is mandatory for "Extra Class" and optional for Class I and II.

ii) Not arranged in regular layers in packages. This presentation is only allowed for Class I and II.

4.12.6 Bulk Packing Containers

In India, mandarin fruits are packed in wooden boxes for long distance shipments/transportation. It is estimated that 8-9 million wooden boxes are required annually to pack 'Nagpur' mandarin fruit produced in central India alone. The common lining material is newspaper in the box with paddy straw as a cushioning. With unavailability of wood over the period of time wooden boards and batons have become thinner (thickness reduced from 5-6 mm to 3-4 mm) resulting in damage to minimum 5-10 per cent boxes during transport. Anandswamy and Venkatsubbiah (1976) tried to improve wooden cases (42x32x29 cm internal dimensions) to hold 110-120 mandarins and also developed ventilated corrugated fibre board boxes (46.5 x 27.5 x 27.5 cm internal dimensions) with 18 kg gross weight for export of 'Coorg' mandarins. For transportation of Kinnows, gunny bags and bamboo baskets were reported to be cheaper than CFB and wooden boxes but losses were higher in bags and baskets (Jain and Chauhan, 1994).

To a limited extent cotton plant stalks (CPS) are utilised as domestic fuel by the rural masses in central India, while the bulk of the stalks are disposed off by burning in the field itself. The material is rich in cellulose and is much similar to hard wood in fibre dimensions. It has been found to be suitable raw material for the preparation of pulp and paper. The stacking strength of the box can be achieved with kraft paper of desired grammage. As a substitute for wooden boxes, corrugated fibre board boxes from commercially available kraft paper and the kraft paper from cotton plant stalks (CPS) have been designed and developed (Ladaniya *et al.*, 1999; Shaikh *et al.*, 2003).

Since the water absorption is a major problem in CFB boxes under high humidity conditions of refrigerated storage, water vapour barrier options such as 1). lamination with PP film, 2).coating with wax, 3).use of bitumen paper and 4). resin coating can be utilized.

The cost of the CFB box has been a major constraint in its wide acceptance by the mandarin industry in India and hence boxes with cost comparable to conventional wooden box have been developed. The one piece regular slotted type cartons cost comparatively less. The cost of CPS kraft paper boxes is also on par with wooden boxes. Corrugated fibre board (CFB) cartons of following specifications have been suitable for packing, transport and storage of 'Nagpur' mandarin fruits. The specifications are as follows: Two piece, 5 ply, full telescopic, laminated with polypropylene film from outside, B flute vertical, 48.5 x 28.5 x 28.5 cm inner dimensions, 50 x 30 x 30 cm outer dimensions, bursting strength 10 kg/cm^2 (5 ply), compression strength 460kg weight of box (at 25-30 °C; 50-60 per cent RH) about 1.10 Kg with cobb value less than 50 or nil. These boxes are most suitable with reference to fruit count and utilization of space in transport vehicle. Number of fruits packed in the box varies with the size of the fruit. The count of fruit in the carton is similar to that of conventional wooden box as given in following Table 19. Boxes with 19.25 kg/cm^2 bursting strength of the board are desirable for export.

Table 19: Fruit Size and Count of 'Nagpur' Mandarin in Conventional Wooden Boxes and CFB Cartons

Wooden Box with Paddy Straw Cushion (45.5 x 32.5 x 32.5cm)			*Telescopic CFB Box with Paper Board Dividers (50x30x30cm)*		
Fruit Size	*Fruit Diameter Average (cm)*	*Count*	*Fruit Size*	*Fruit Diameter (cm)*	*Count*
Extra large	7.55	80-120	Extra large	7.31-8.00 (Avg. 7.65)	96-104
Large	7.00	120-140	Large	6.71-7.30 (Avg. 7.00)	112-120
Medium	6.40	150-180	Medium	6.11-6.70 (Avg. 6.40)	140-175
Small	5.80	170-200	Small	5.50-6.10 (Avg. 5.80)	200
Very small	5.00	225-270	Very small	4.70-5.49 (Avg. 5.00)	225-250

Source: Ladaniya *et al.*, 1999.

Aeration vents at 4-5 per cent of side areas punched as 4 long slits (9.5 cm x 1.75 cm) and 1.65 per cent of area punched at ends with one slit (8 x 1.7 cm) as a handling slot are sufficient. Ventilation up to 6 per cent of the side areas of the box provided better aeration during precooling in shipping containers.

Nearly 40 to 50 per cent of 'Nagpur' mandarin fruit produced in Central India is packed in conventional wooden boxes (48x 35 x 35 cm) for dispatch to distant markets either from orchard or assembly markets. Earlier gunny bags or bamboo baskets with paddy straw were used for packing of Kinnows produced in Punjab and Rajasthan and transported by truck (Chauhan *et al.*, 1987; Chopra and Dhar, 1990). Since 1995-96 CFB cartons are increasingly being used for packing of Kinnow fruit as the fruit is waxed in modern packing houses. Presently more than 80 per cent Kinnow mandarin fruit from Rajasthan and Punjab states (India), is packed in 10 kg capacity boxes (48-84 per carton count; carton size 45 x 23 x 18 cm) for export (Table 20). Ten kg capacity carton was also used by some exporters for 'Nagpur' mandarin.

Table 20: Size and Number of Kinnows in 10 kg Box

'Kinnow' Size (mm)	*Number of Fruit in 10 kg Box*
60 – 64	84
65 – 69	72
70 – 72	60
72-74	54
75-79	51
80-85	48

For export of Nagpur mandarin to countries in Middle–East, fruit of specified grade and weight need to be packed in CFB cartons (NHB, 2011). The packaging and other details are given in Table 21:

Table 21: Fruit Specifications for Export

Colour	Light orange
Fruit weight	150-170 g for 65-70 mm fruit,90-100g for 45-50 mm fruit
Size (diameter)	65-70 mm to be packed in 7 kg box
	45-50 mm to be packed in 2.5 kg box
Temperature for shipping	5-7 °C
Shipping	By sea (Reefer)

The CFB cartons used for mandarins in California are two piece full telescopic type half slotted with side vents. The dimensions are 41.6cm (L) x 27.2cm (W) x 19.0-21.6cm (H) and 41.6cm (L) x 27.2cm (W) x 26cm (H) with volume nearly 28 lit. The fruit count is : Jumbo – 100 count (6.35 – 6.98cm diameter), Large – 135 count (5.72 – 6.35cm diameter), Medium – 180 count (5.08 – 5.72cm diameter. Small – 230 count (4.44 – 5.08cm).

4.13 Mango (*Mangifera indica*)

4.13.1 Minimum Requirements

In all classes, subject to the special provisions for each class and the tolerances allowed, the mangoes must be:

i) Whole;

ii) Sound, produce affected by rotting or deterioration such as to make it unfit for consumption is excluded;

iii) Clean, practically free of any visible foreign matter;

iv) Practically free of damage caused by pests;

v) Free of abnormal external moisture, excluding condensation following removal from cold storage;

vi) Free of any foreign smell and/or taste;

vii) Firm;

viii) Fresh in appearance;

ix) Free of damage caused by low temperatures;

x) Free of black necrotic stains or trails;

xi) Free of marked bruising; and

xii) Sufficiently developed and display satisfactory ripeness.

When a peduncle is present, it shall not be no longer than 1.0 cm.

Mangoes shall comply with the residue levels of heavy metals, pesticides and other food safety parameters as laid down by the Codex Alimentarius Commission for exports.

4.13.2 Grade Designation

Mangoes are classified in to three grades/classes depending on fruit quality (Table 22).

4.13.3 Provisions Concerning Sizing

Size is determined by the weight of the fruit, in accordance with the Table 23.

4.13.4 Size Tolerance

For all grades, 10 per cent by number or weight of mangoes in each package are permitted to be outside (above or below) the group size range by 50 per cent of the maximum permissible difference for the group. In the smallest size range, mangoes must not be less than 90 gms. and for those in the largest size range a maximum of 925 grams applies, as follows (Table 24).

4.13.5 Packaging

i) At the time of packing, development and condition of the mangoes must be such as to enable them to ensure a continuation of the maturation until they reach appropriate degree of maturity corresponding to the varietal characteristics.

Table 22: Grade Criteria for Mangoes

Grade Designation	*Grade Requirements*	*Grade Tolerances*
Extra class	Mangoes must be of superior quality. They must be characteristic of the variety. They must be free of defects, with the exception of very slight superficial defects, provided these do not affect the general appearances of the produce, the quality, the keeping quality and presentation in the package.	5 per cent by number or weight of mangoes not satisfying the requirements for the grade, but meeting those of Class I or exceptionally, coming within the tolerances of that grade.
Class – I	Mangoes must be of good quality. They must be characteristic of the variety. Mangoes may have following slight defects, provided these do not affect the general appearance of the produce, the quality, the keeping quality and presentation in the package. ☆ Slight defects in shape; ☆ Slight skin defects due to rubbing or sunburn, suberized stains due to resin exudation (elongated trails included) and healed bruises not exceeding 2,3,4,5 sq. cm. for size groups A, B, C, D, respectively	10 per cent by number or weight of mangoes not satisfying the requirements for the grade, but meeting those of Class II grade or, exceptionally coming within the tolerances of that grade.
Class – II	This grade includes mangoes which do not qualify for inclusion in the higher grades, but satisfy the minimum requirements. Mangoes may have following defects, provided they retain their essential characteristics as regards the quality, keeping quality and presentation. ☆ Defects in shape, ☆ Slight skin defects due to rubbing or sunburn, suberized stains due to resin exudation (elongated trails included) and healed bruises not exceeding 4,5,6,7 sq. cm. for size groups A, B, C, D respectively.	10 per cent by number or weight of mangoes not satisfying the requirements of the grade, but meeting the minimum requirements.

In Class I and Class II, scattered suberized rusty lenticels, as well as yellowing of green varieties due to exposure to direct sunlight, not exceeding 40 per cent of the surface and not showing any signs of necrosis are allowed.

Table 23: Mango Fruit Size and Code

Size Code	*Weight (g)*	*Maximum Permissible Size difference between the Fruit within the Package (g)*
A	100-200	50
B	201-350	75
C	351-550	100
D	551-800	125

Table 24: Size Range, Permissible Size difference

Size Code	*Normal Size Range*	*Permissible Size Range (10 per cent of fruit/package exceeding the normal size range)*	*Maximum Permissible difference between Fruit in each Package*
A	100-200	90-220	50.00
B	201-350	18-425	112.50
C	351-550	251-650	150.00
D	551-800	429-925	187.50

ii) Fruit shall be firm enough to withstand transport and handling, and to arrive in satisfactory condition at the place of destination.

For export purpose, packaging of mangoes is done in corrugated fibre board boxes having outer ply of duplex board as detailed below in the Table 25. Only packing material of food grade should be used for packing mangoes at the pack-house facility and the package boxes having dimensions of 370 X 275 X 90 mm should be used for packing export mangoes, as approved by US Department of Agriculture- Animal and Plant Health Inspection Service (USDA-APHIS), Each package must be either preprinted or affixed with a specified label, which is duly approved by the US Department of Agriculture- Animal and Plant Health Inspection Service (USDA-APHIS). The label should be appropriately marked/stamped on left-half side indicating Production Unit Code Number (PUC), Packinghouse Code Number (PHC), Date of Packing, and Lot Number.

Table 25: Specifications for Corrugated Fibre Board (CFB) Boxes for Packing Mango for Export

Characteristics	*Requirements*			
	Ring and Flap (4 kg)	*Full Telescopic (4 kg)*	*Ring Flap (8 kg)*	*Full Telescopic (8 kg)*
i) Material used	3 Ply Corrugated fibre board	3 Ply Corrugated fibre board	5 Ply Corrugated fibre board	5 Ply Corrugated fibre board
ii) Grammage (g/m sq), Min (outer to inner)	*230/140/140	*230/140/140	*230/140/140	*230/140/140
iii) Bursting strength kg/cm sq, Min	6.50	6.50	10.50	6.50
iv) Puncture resistance, ozs inches/tear inch Min	110.00	110.00	280.00	110.00
v) Compression strength, kgf, Min	275.00	275.00	250.00	250.00
vi) Cobb (30 minutes) g/m sq, Max	130.00	130.00	130.00	130.00

* Outer ply of duplex board ; Source: Postharvest Manual for Export of Mangoes, APEDA, New Delhi.

While packing suitable holders (polystyrene foam individual fruit holders and bubble sheets etc need to be used to avoid any injury to the fruit. Recommendation of APEDA for packaging of mango for export purpose are as follows: 1)Each individual fruit of mango shall be enclosed in a clean, white, soft, expandable and netted type polystyrene sleeve to prevent bruising before packing in a box. 2) The mangoes must be packed in insect-proof boxes. If ventilated boxes are used, all the ventilator openings of the box should be covered with insect-proof screen of minimum 30 mesh and all the sides of box should be sealed with adhesive tape to prevent any entry of pests. 3)The materials used inside the package must be of food grade, new, clean, and of a quality such as to avoid causing any external or internal damage to the produce. 4)The use of materials, particularly of paper or stamps bearing trade specifications is allowed, provided the printing or labeling has been done with non-toxic ink or glue(APEDA, 2007).

For domestic market, usually mangoes are packed in wooden boxes, details of which are given below (Table 26). However, Alphonso and Kesar are packed in CFB boxes.

Table 26: Specifications of Wooden Boxes for Domestic Market

Type of Carton	*Inner Dimension (cm)*	*Capacity (kgs)*
Wooden crates	45 × 30 × 30 (Ratnagiri) 21.6 × 21.6 × 42 (Malihabad)	16-18 10-11

For export of mangoes to countries in Europe and Middle –East, fruit of specified weight need to be packed in CFB cartons (NHB, 2011). Variety and other details are given in Table 27.

4.13.6 MAP/Shrink Wrapping

Green Alphonso mangoes wrapped with heat-shrinkable film were stored up to 5 weeks at 8 °C in acceptable condition (Figure 5). Fruits ripened normally at room temperature during one week ripening period and further had shelf life of one week (Sudhakar Rao, Personal communication, 2013).This indicated that shelf life of Alphonso mangoes can be up to 7 weeks by shrink wrapping.

Figure 5: Shrink Wrapped Green Alphonso Mangoes after Ripening

Table 27: Specifications for Export of Mango to Europe and Middle- East

Variety	Weight (g)			Packing (kg)			Shipment			Storage Temperature
	Netherlands/ Germany	U.K.	Middle East	Netherlands/ Germany	U.K.	Middle East	Netherlands/ Germany	U.K.	Middle East	
Alphonso	250-300	250-300	250-250	1 Doz/2.5	1 Doz/2.5	1 Doz/2.5	By Air	By Air	By Sea	13 °C
Kesar	225-250	225-250	200-250	1 Doz/2.5	1 Doz/2.5	1 Doz/2.5	By Air	By Air	By Sea	13 °C

4.14 Papaya (*Carica papaya*)

4.14.1 Minimum Requirements

Papaya shall be whole, fresh in appearance, firm, sound, clean, free of any type of damage or smell.

The peduncle, if present, should not exceed a length of 1 cm.

Papaya shall comply with the residue levels of heavy metals, pesticides and other food safety parameters as laid down by the Codex Alimentarius Commission for exports.

4.14.2 Grade Designation

Papayas are graded in three grades:

i) Extra class

ii) Class – I

iii) Class – II

Fruit of Extra class must be of superior quality without surface defects. The tolerance limit for this class is 5 per cent by number or weight of fruit not satisfying the requirements of the grade, but meeting those of Class I or, exceptionally, coming within the tolerances of that grade.

For Class I, papayas in this class must be of good quality with surface defects covering 10 per cent area but not affecting pulp. Tolerance is 10 per cent by number or weight of fruit not satisfying the requirements of the grade, but meeting those of next class. For Class –II, papayas must meet minimum requirement of quality and surface defects covering 15 per cent area are allowed but not affecting pulp. Tolerance is 10 per cent by number or weight of fruit not satisfying the requirements of the grade but meeting the minimum requirements can be included.

4.14.3 Provisions Concerning Sizing

Size is determined by the weight of the fruit with a minimum weight of 200 g. in accordance with the Table 28.

Table 28: Papaya Size Code and Corresponding Weight

Size Code	*Weight (g)*
A	200-300
B	301-400
C	401-500
D	501-600
E	601-700
F	701-800
G	801-1100
H	1101-1500
I	1501-2000
J	>2001

4.14.4 Size Tolerance

For all grades, 10 per cent by number or weight of papayas corresponding to the size immediately above and/or below that indicated on the package, with a minimum of 190 g. for those papayas packed in the smallest size range.

4.14.5 Packaging

i) At the time of packing, papayas must have been carefully picked and have reached an appropriate degree of development and ripeness, account being taken of the characteristics of the variety and/or commercial type and to the area in which they are grown.

ii) The development and condition of the papayas must be such as to enable them to withstand transport and handling, and to arrive in satisfactory condition at the place of destination.

a) Shrink Wrapping

Different types of polymeric films *viz.*, LDPE, PP, Pebax® and storage temperatures *viz.* 7, 13 °C and 27-32 °C have been tried for individual fruit packing of green "Solo" papaya. MAP of individual papaya fruits with Pebax® or LDPE film could extend the storage life and maintain the quality up to two weeks over one week in non-packed fruits under ambient conditions (27-32 °C). At low temperature (13 °C), the storage life of papaya could be extended to one month when packed in LDPE or Pebax® film without any chilling injury (CI) symptoms and the fruits ripened normally in one week under controlled conditions (20 ±1 °C and 75-80 per cent RH).The fruits stored at 7 °C failed to ripen normally and showed severe CI symptoms and disease incidence (Sukhvinder Pal Singh, 2003).

For export of papaya fruit to Europe and Middle-East, packaging and fruit specifications have been recommended by National Horticulture Board (NHB, 2011) as given in Table 29.

4.15 Pear (*Pyrus pyrifolia*)

4.15.1 Minimum Requirements

Fruit should be wholesome, sound, clean and free from defects

4.15.2 Grade Designation

Pears are graded in three grades.

i) Extra class

ii) Class – I

iii) Class – II

Fruit of Extra class must be of superior quality without surface defects. The tolerance limit for this class is 5 per cent by number or weight of fruit not satisfying the requirements of the grade, but meeting those of Class I or, exceptionally, coming within the tolerances of that grade.

For Class I, pears must be of good quality with surface defects covering 10 per cent area but not affecting pulp. Tolerance is 10 per cent by number or weight of fruit

Table 29: Specifications for Export of Papaya

Variety	Weight		Packing (kg)		Shipment	Storage Temperature
	Europe	Middle East	Europe	Middle East		
Solo	400-500 g/fruit		5	8	By sea	10-13 °C
Taiwan 785	400-500 g/fruit	1-1.25 kg/Fruit	5	8	By sea	10-13 °C
Taiwan 786	400-500 g/fruit	1-1.25 kg/Fruit	5	8	By sea	10-13 °C

not satisfying the requirements of the class/grade, but meeting those of next class. For Class –II, fruit must meet minimum requirement of quality and surface defects covering 15 per cent area are allowed but not affecting pulp. Tolerance is 10 per cent by number or weight of fruit not satisfying the requirements of the grade but meeting the minimum requirements.

4.15.3 Provisions Concerning Sizing

i) Size is determined by maximum diameter of the equatorial section.

The minimum size by diameter is required for all grades as follows:

Extra Class: 60 mm; Class-I: 55mm ; Class-II: 50 mm

The difference in diameter between fruit in the same package shall be limited to:

i) 5 mm for Extra Class, Class-I and Class-II fruit packed in rows and layers.

ii) 10 mm for Class-I, Class-II fruit packed in bulk in the package.

4.15.4 Size Tolerance

For all classes, 10 per cent by number or weight of fruit not corresponding to the size with a maximum variation of 5mm below the minimum.

4.15.5 Packaging

i) At the time of packing, development and condition of the pears must be such as to enable them to continue their maturing process and to reach the appropriate stage of physiological maturity corresponding to the particular characteristics of the variety.

ii) Fruit shall be able to withstand transport and handling and be able to arrive in satisfactory condition at the place of destination.

iii) The contents of each package must be uniform and contain only of the same origin, variety, quality, size and same degree of ripeness.

iv) In the case of, fruit colour shall be uniform.

Pears are packed in corrugated fibre-board boxes of required strength for bulk packaging and transport. Paper pulp trays are used for fruit of "Extra Class". Consumer or retail packaging is done in small trays and wrapped with heat shrinkable film.

4.16 Pineapple (*Ananas comosus*)

4.16.1 Minimum Requirements

Pineapples must have been carefully picked and have reached an appropriate degree of development and ripeness in accordance with criteria proper to the variety and/or commercial type and to the area in which they are grown.

When peduncle is present, it shall be no longer than 2.0 cm. and the cut must be transversal, straight and clean. The fruit must be physiologically ripe *i.e.* without evidence of unripeness (opaque, flavourless, exceedingly porus flesh) or over ripeness (exceedingly translucent or fermented flesh). Trimming consists in tearing some leaves off the top of the crown.

4.16.2 Grade Designation

Pineapple are graded in three grades:

i) Extra class

ii) Class – I

iii) Class – II

Total soluble solids in Brix degrees (Minimum) shall be 12 in all the classes.

Fruit of "Extra class" must be of superior quality without surface defects. The tolerance limit for this class is 5 per cent by number or weight of fruit not satisfying the requirements of the grade, but meeting those of Class I or, exceptionally, coming within the tolerances of that grade. The crown, if present, shall be simple and straight with no sprouts and shall be between 50 and 150 per cent of the length of the fruit for pineapples with trimmed or untrimmed crowns.

For "Class I", pineapple in this class must be of good quality with surface defects covering 10 per cent surface area but not affecting pulp. Grade tolerance is 10 per cent by number or weight of fruit not satisfying the requirements of the grade, but meeting those of next class. The crown, if present shall be simple and straight and slightly curved with no sprouts, and shall be between 50 and 150 per cent of the length of the fruit for pineapples with trimmed or untrimmed crowns.

For "Class –II", pineapple must meet minimum requirement of quality and surface defects covering 15 per cent area are allowed but not affecting pulp. Grade tolerance is 10 per cent by number or weight of fruit not satisfying the requirements of the grade but meeting the minimum requirements can be included. The crown, if present, shall be simple or double and straight or slightly curved, with no sprouts.

4.16.3 Provisions Concerning Sizing

Size is determined by the average weight of the fruit with a minimum weight of 700 g. except for small size varieties (such as victoria and queen) which can have a minimum weight of 250 g. in accordance with the Table 30.

Table 30: Pineapple Fruit Size, Code and Weight

Size Code	*Average Weight (g) (+ or – 12 per cent)*	
	With Crown	*Without Crown*
A	2750	2280
B	2300	1910
C	1900	1580
D	1600	1330
E	1400	1160
F	1200	1000
G	1000	830
H	700	560
I	< 700	< 560

4.16.4 Size Tolerance

For all classes, 10 per cent by number or weight of pineapples corresponding to the size immediately above and/or below that indicated on the package.

4.16.5 Packaging

At the time of packing, development and condition of the pineapples must be such as to enable them to withstand transport and handling, and enable fruit to arrive in satisfactory condition at the place of destination.

Fruits are packed in printed corrugated fibre–board boxes of required strength. Ventilation is provided to boxes. Fruit shall not be over-ripe at the time of packing.

4.17 Plum *(Prunus domestica)*

4.17.1 Minimum Requirements

Plums must have been carefully picked and have reached an appropriate degree ofdevelopment and ripeness in accordance with criteria proper to the variety and/or commercial type and to the area in which they are grown.

4.17.2 Grade Designation

Plums are classified in three classes depending on fruit quality (Table 31).

Table 31: Grade Designation for Plums

Grade Designation	*Grade Requirements*	*Grade Tolerance*
Extra class	Plums shall be of superior quality. They must have the shape, development and colouring typical of the variety. They must be ☆ Free from defects; ☆ Practically covered by bloom, according to the variety; ☆ Of firm flesh.	5 per cent by number or weight of Plums not satisfying the requirements of the grade, but meeting those of Class I grade or, exceptionally, coming within the tolerances for that class.
Class – I	Plums must be of good quality. They must be characteristics of the variety and/or commercial type. The following slight defects of the fingers, however, may be allowed, provided these do not affect the general appearance of the produce, quality, the keeping quality and presentation in the package. ☆ A slight defect in shape, ☆ A slight defect in colouring; ☆ A slight defect in development; ☆ Skin defects of elongated shape that must not exceed in length one-third of the maximum diameter of the fruit; ☆ Other skin defect, of which total area affected must not exceed 5 per cent of the whole surface;	10 per cent by number or weight of Plums not satisfying the requirements of the grade, but meeting those of Class II grade or, exceptionally, coming within the tolerances of that grade.

Contd...

Table 31–Contd...

Grade Designation	*Grade Requirements*	*Grade Tolerance*
	☆ The stem may be damaged or missing, provided that there is no risk of the fruit rotting in consequence.	
Class – II	Plums which do not qualify for inclusion in the higher classes, but satisfy the minimum requirements. ☆ Defects in shape, development and colouring are allowed provided that the plums retain their characteristics. ☆ Skin defects not liable to impair the external appearance of the fruits or its keeping quality are allowed provided that they do not exceed 10 per cent of the whole surface.	10 per cent by number or weight of Plums not satisfying the requirements for the class but meeting the minimum requirements.

4.17.3 Provisions Concerning Sizing

Sizing is determined by the maximum diameter of the equatorial section (Table 32).

Table 32: Plum Fruit Grade and Corresponding Fruit Diameter

Grade	*Large Fruit Varieties (Min.) (mm)*	*Other Varieties (mm)*	*Mirabelles and Damsons (mm)*
Extra class	40	28	20
Class I	35	28	20
Class II	30	25	17

4.17.4 Size Tolerance

For all classes, 10 per cent by number or weight deviating from the minimum size or the size stated on the packages the deviation not exceeding 3 mm. above or below.

4.17.5 Packaging

i) At the time of packing, plums must be sufficiently developed and display satisfactory ripeness. The development and condition of the plums must be such as to enable them to withstand transport and handling and to arrive in satisfactory condition at the place of destination.

ii) Plums are prone to damage during transport and hence boxes and cushioning shall be such as to provide necessary protection during long distance transport.

4.18 Pomegranate (*Punica granatum*)

4.18.1 Minimum Requirements

Pomegranates must be carefully picked and have reached an appropriate degree of development and ripeness in accordance with criteria proper to the variety.

4.18.2 Grade Designation

Pomegranate are graded in three grades:

i) Extra class

ii) Class – I

iii) Class – II

Fruit of Extra class must be of superior quality without surface defects. The tolerance limit for this class is 5 per cent by number or weight of fruit not satisfying the requirements of the grade, but meeting those of Class I or, exceptionally, coming within the tolerances of that grade.

For Class I, pomegranate must be of good quality with surface defects covering 10 per cent area but not affecting pulp. Tolerance is 10 per cent by number or weight of fruit not satisfying the requirements of the grade, but meeting those of next class. For Class –II, fruit must meet minimum requirement of quality and surface defects covering 15 per cent area are allowed but not affecting pulp. Tolerance is 10 per cent by number or weight of fruit not satisfying the requirements of the grade but meeting the minimum requirements can be included.

4.18.3 Provisions Concerning Sizing

Size is determined by the weight or maximum diameter of the equatorial section of the fruit, in accordance with the Table 33.

Table 33: Pomegranate Fruit Size Code, Weight and Diameter

Size Code	*Weight (in grams) (Minimum)*	*Diameter in mm (Minimum)*
A	400	90
B	350	80
C	300	70
D	250	60
E	200	50

4.18.4 Size Tolerance

i) For all grades, 10 per cent by number or weight of pomegranate corresponding to the size immediately above and/or below than indicated on the package.

ii) The maximum size range of 8 mm. between fruit in each package is permitted

4.18.5 Packaging

At the time of packing, development and condition of the fruit must be such as to enable them to withstand transport and handling and to arrive insatisfactory condition at the place of destination.

The main problems associated with prolonged storage of pomegranate fruit are weight loss and shrinkage, decay development, appearance of skin blemishes, especially scald, and impaired quality and taste. Artes *et al.* (2000) stored Spanish 'Mollar de Elche' sweet pomegranates at 5 °C for 12 wks in un-perforated polypropylene (UPP) film of 25 µm thickness in modified atmosphere packaging (MAP). MAP strongly reduced water loss and chilling injuries without incidence of decay. Porat *et al.* (2008) reported simple and low-cost MAP technology to alleviate storage problems and maintain fruit quality for up to 3-4 months after harvest. For moderate storage up to 2 months, packing of fruit in Xtend® Easy-Tear® bags in small 4-5 kg retail cartons has been recommended. This type of bag includes a notch, which allows the final user to easily tear and remove the top part of the bag for retail display. The main advantage of packing the fruit in Xtend® Easy-Tear® bags in export cartons is maintenance of fruit quality and reduction in shrinkage and scald development both during the storage period and subsequent marketing. For prolonged storage up to 3-4 months after harvest, packing of fruit in 20 kg bulk bags in plastic crates or in 80 kg bulk bags suitable for storing fruit in harvest bins has been suggested. For retail storage, repacking of fruit in to Xtend® Easy-Tear® bags is necessary.

Shelf life of 'Bhagwa' pomegranate was extended up to 3 months with heat–shrinkable film wrapping and storage at 8 °C. Under ambient storage condition, shrink wrapping extended storage life of fruit up to one month (Figures 6a and b) (Sudhakar Rao, 2013, personal communication).

For export of pomegranate to some European and Gulf countries, packaging and fruit specifications have been suggested by NHB, (2011) as per preference of countries for the fruit (Table 34). Fruit shall have pleasant flavour and aroma and higher TSS (16-17 Degree Brix) is preferred. Fruit shall be free from any type of blemish on the surface and uniformity of size and shape in the package is of paramount importance.

4.19 Sapota (*Achras sapota*)

4.19.1 Minimum Requirements

Sapota shall have been carefully picked and have reached an appropriate degree of development and displayed satisfactory ripeness in accordance with criteria proper to the variety and to the area in which they are grown. The development and condition of fruit must be such as to enable them to withstand transport and handling, and to arrive in satisfactory condition at the place of destination. Sapota shall not be very soft, shriveled or water soaked.

Sapota should comply with the residue levels of heavy metals, pesticides and other food safety parameters as laid down by the Codex Alimentations Commission for exports.

Table 34: Fruit and Packaging Specifications for Export of Pomegranate

Variety	*Colour*	*Weight (g)*			*Packing (kg)*			*Shipment*			*Storage Temperature*
		Netherlands/ Germany	*U.K.*	*Middle East*	*Netherlands/ Germany*	*U.K.*	*Middle East*	*Netherlands/ Germany*	*U.K.*	*Middle East*	
Ganesh	Red	250-300	250-300	300-450	3	3	5	By Sea	By Sea	By Sea	5 °C
Arakta, Mrudula	Deep Red	200-250	200-250	200-250	3	3	5	By Sea	By Sea	By Sea	5 °C

Figure 6: Shrink Wrapped Pomegranate
(a) Ambient Condition (Above) (b) Refrigerated Condition (Below)

4.19.2 Grade Designation

Sapota are graded in three grades:

i) Extra class

ii) Class – I

iii) Class – II

Fruit of "Extra class" must be of superior quality without surface defects. The tolerance limit for this class is 5 per cent by number or weight of fruit not satisfying the requirements of the grade, but meeting those of Class-I or, exceptionally, coming within the tolerances of that grade.

For Class I, fruit must be of good quality with surface defects covering 10 per cent area but not affecting pulp. Grade tolerance is 10 per cent by number or weight of fruit not satisfying the requirements of the grade, but meeting those of next class. For "Class –II", fruit must meet minimum requirement of quality and surface defects covering 15 per cent area are allowed but not affecting pulp. Grade tolerance is 10 per cent by number or weight of fruit not satisfying the requirements of the grade but meeting the minimum requirements can be included.

4.19.3 Provisions Concerning Sizing

Size is determined by the weight of the fruit (Table 35).

Table 35: Sapota Fruit Grade and Corresponding Weight

Class	*Weight (g)*
Extra class	101 and above
Class – I	100
Class – II	80

The difference in weight between the largest and the smallest fruit in any one package must not exceed:

i) 15 gms. for fruits weighing less than 100 gms.

ii) 20 gms. for fruits weighing between 100 gms. and above.

4.19.4 Size Tolerance

For all grades, 10 per cent by number or weight of sapota not conforming to the minimum weight and or the size range specified.

As a practice of sapota marketing in Gujarat, fruit grading is mainly based on size and shape of the fruits. Sizes are only three *viz.* large, medium and small. The study by Varshney *et al.* (2005) revealed that small farmers grade sapota manually into 2 to 3 sizes before marketing. The farmers' cooperative marketing societies do the grading and packaging operations. The prices of fruits are paid to the growers on the basis of fruit count per 10 kg. Traders also grade the fruits manually and supply them to meet the requirement of the cities at short distances.

4.19.5 Packaging

While the packaging of fruits is done in ventilated corrugated fiberboard boxes by cooperative societies and traders, gunny bags are used by the farmers in Gujarat for packaging of sapota fruits (Varshney *et al.*, 2005). Fruits are packed in cardboard boxes of 10 kg. capacity with rice straw as padding material and ethylene absorbents. Bamboo baskets with banana leaves as lining material and covered with net are used for carrying the produce from farm to local market. In some cases, produce is packed in wooden containers and transported to the urban areas for sale. Wooden boxes prevent damage of ripened fruits and fetch better price in urban areas. Sapota are also packed in corrugated fibre board boxes with paper clippings as cushioning material (Figure 7) (Sudhakar Rao, personal communication, 2013).

Figure 7: Sapota Fruit Packed in Corrugated Fibre Board Boxes

4.19.6 MAP

Based on the respiration rate and permeability of the packaging material, the suitable packaging films reported for the modified atmospheric packaging of sapota fruit were low density Polyethylene, Polyvinyl chloride, Polypropylene and Polystyrene film (Dash, *et al.*, 2012). Saran and polyester films were found to be unsuitable for MAP storage of the sapota fruit. These researchers have developed a model that provides an effective way for selection of the proper packaging material for sapota fruit MAP and this model can also be employed for designing packaging films for other fruits also as per the respiration rate of the fruit.

For export by sea rout to countries in Middle-East, packaging of Cricket Ball and Kalipatti sapota fruit in 3 kg carton and temperature of 15-17 °C has been suggested by National Horticulture Board (NHB, 2011).

4.20 Strawberry (*Fragaria* spp.)

4.20.1 Minimum Requirements

Strawberries should be sound, clean and free from any defect when harvested.

4.20.2 Grade Designation

Strawberries are graded in three grades:

i) Extra class

ii) Class – I

iii) Class – II

Fruit of Extra class must be of superior quality.

4.20.3 Provisions Concerning Sizing

Size is determined by the maximum diameter of the equatorial section of the fruit.

Strawberries must have following minimum sizes.

Extra Class : 28 mm and Class I and II: 18 mm.

4.20.4 Size Tolerance

For all grades, 10 per cent by number or weight of strawberries not conforming to the minimum size provided, however, that the diameter is not less than 16 mm.

4.20.5 Packaging

Strawberries have very short shelf life and need to be carefully handled and packed. Generally for retailing fruit is packed in plastic punnets holding 9-12 fruits (Figure 8).

For bulk transport small punnets can be packed in master containers. Following care shall be taken:

i) For packing, strawberries must have been carefully picked and have reached an appropriate degree of development and display satisfactory ripeness in accordance with criteria proper to the variety and to the area in which they are grown.

ii) The development and condition of the strawberries must be such as to enable them to withstand transport and handling, and enable fruit to arrive in satisfactory condition at the place of destination.

Figure 8: Strawberry Packed in Plastic Punnet

iii) Strawberries in the 'Extra Class' must be well presented.

iv) Packages must be free from all foreign matter.

4.21 Sweet Orange (*Citrus sinensis*)

4.21.1 Minimum Requirements

Oranges shall be fresh in appearance, wholesome, with characteristic shape of the variety, sound, clean, free of any visible foreign matter, free of damage or defect or smell or condensed moisture caused by low temperature or high temperature. The stalk shall be removed close to the fruit keeping the button intact. Oranges shall comply with the residue levels of heavy metals, Pesticides and other food safety parameters as laid down by the Codex Alimentarius Commission for exports. Oranges shall have minimum diameter of 53 mm.

4.21.2 Grade Designation

Sweet oranges are graded/classified in to three classes depending on quality (Table 36).

4.21.3 Provision Concerning Sizing

Size is determined by the maximum diameter of the equatorial section of the fruit in accordance with the Table 37.

4.21.4 Size Tolerance

For all classes, 10 per cent by number of oranges corresponding to the size immediately below and/or above that mentioned on the packages, provided the maximum difference between the smallest and largest fruits does not exceed the following maxima (Table 38).

Table 36: Grade Criteria for Oranges

Grade Designation	*Grade Requirements*	*Grade Tolerance*
Extra class	Extra class oranges shall be of superior quality. They shall be characteristic of the variety and/or commercial type in shape, external appearance, development and colouring. They must be free of defects, with the exception of very slight superficial defects provided these do not affect the general appearance of the produce, the quality, the keeping quality and presentation in the package; ✰ Shall be uniform in colour; ✰ Shall have minimum diameter of 70 mm.	5 per cent by number or weight of Oranges not satisfying the requirements of the grade, but meeting those of Class I grade or, exceptionally, coming within the tolerances of that grade.
Class – I	Grapes Class I oranges shall be of good quality. They must be characteristic of the variety or commercial type (in shape, external appearance, development and colouring). Following slight defects may be there, provided these do not affect the general appearance of the produce, the quality, the keeping quality and presentation in the package; ✰ Slight defects in colouring, ✰ Slight defects in shape, ✰ Slight skin defects occurring during the formation of the fruit, such assilver scurf, russeting etc., ✰ Slight healed defects due to mechanical cause such as hail damage, rubbing, damage from handling.	The defects shall not, in any case, affect the pulp of the fruit. 10 per cent by number or weight of Oranges not satisfying the requirements of the grade, but meeting those of Class II or, exceptionally, coming within the tolerances of that grade.
Class – II	This grade includes Oranges which do not qualify for inclusion in the higher grades, but satisfy the minimum requirements. Following defects may be there, provided the Oranges retain their essential characteristics as regards the quality, the keeping quality and presentation in the package. ✰ Defects in shape; ✰ Defects in colouring; ✰ Skin defects occurring during the formation of the fruit, such as silver scruffs, russeting etc.; ✰ Skin defects, healed defects due to mechanical cause such as hail damage, rubbing, damage from handling; ✰ Superficial healed skin alteration; ✰ Rough skin.	The defects shall not, in any case, affect the pulp of the fruit. 10 per cent by number or weight of Oranges not satisfying ther equirements of the grade, but meeting the minimum requirements. Within this tolerance, a maximum of 5 per cent may be fruits showing slight superficial unhealed damage, drycuts or soft and shrivelled fruit.

Table 37: Size Code and Fruit Diameter for Pranges

Size Code	*Diameter (mm)*
A	53-60
B	61-68
C	69-76
D	77-84
E	85-92
F	93-100
G	Above 100

Table 38: Sweet Orange Size Code and Permitted difference in Diameter

Size code	*Maximum difference between Fruit in the Same Package (mm)*
A,B,C and D	7
E, F and G	10

4.21.5 Mechanized Sizing and Grading

Sizing of sweet oranges both on the basis of weight and dimensions is better as compared to only weight-sizing or only dimension-sizing. Disorders like granulation and frost damage may create problem for correct determination of sizing as the fruit lose juice and weight in these disorders but the fruit size (dimensions) remains same. The microprocessor based (computer aided) modern sizers can weigh the individual fruit in the specially designed cups. These cups weigh the fruit and accordingly drop it on cross belts below the cups. If fruit is of higher weight then it moves further to next cup. Fruit with highest weight falls last. Weight sizers may be useful in case of oranges and grapefruit.

Dimension sizer has to be designed in order to measure and segregate fruit on the basis of 2 or 4 contact points. Measurement on the basis of 4 contact points is most accurate since diameter is usually greater or smaller in citrus fruit as compared to height or length. When fruit is in contact at four points it is called as volumetric sizer. These are best suitable for lemons which are oblong to ovate in shape with greater length and relatively smaller diameter. Sizing of fruits should be fairly accurate for it may affect volume and lead to problems in uniformity of packaging. The packers will face problem maintaining uniformity of fruit in package and that could slow their speed in searching right sized fruit.

The 2 point contact sizers can be of two types. In roller sizers, singulated fruit is dropped on the divergent rollers which are positioned with increasing spacing. As the fruit moves from the beginning to the end of the roller, small fruit falls on the pull-out belt below the roller while larger fruit continues along the rollers until appropriate spacing between the roller allow the fruit to drop. The rollers can be adjusted for different sizes of the citrus. In another method fruit are conveyed to different size openings, the small opening first and increasing gradually to separate small, medium, big and large sizes. These sizers have been fairly appropriate for oranges and mandarin with near spherical shape. In case of flat shaped fruit such as mandarin/ tangerine with greater diameter and smaller height (length between stem and stylar-end) the sizing could be erroneous and would solely depend on orientation of fruit at contact points.

The divergent roller type sizers or rotating disc type dimension sizers (with 2 contact points) have given fairly good results in case of Kinnow and 'Nagpur' mandarin, respectively provided that contact takes place at equatorial diameter. The rotation speed of the disc has to be adjusted according to the feeding of fruit or else it may lead to congestion and mixing of grades. At the edge of rotating disc, the opening for different sizes of fruit is given which is adjustable. Small fruit is segregated first followed by medium sized and large fruit. Reduction of speed of rotary disc of the sizer from 10 to 8.5 rpm resulted in satisfactory sizing at the feeding rate of 1 tonne/ hr. in case of 'Nagpur' mandarin. This type of sizer could segregate Mosambi orange fruit on dimension basis at two points with more than 90 per cent accuracy in sizing. There was no injury to fruit in mechanical sizing. Divergent roller grader with capacity to grade 400 kg acid lime (*C. aurantifolia* Swingle) fruit per hour can be operated manually also for small scale operations. A similar grader with a capacity of 200-250 kg/hr is reported to segregate Kinnow mandarin fruits on the basis of diameter.

More accurate electronic sizers using optical technology are now available which segregate on the basis of the dimensions and shape. These sizers use light beams and on the basis of blockage of light determine the shape and size. The singulation and orientation of fruit when light is focused on it is very important for accuracy of the sizing. In Japan, fully automated opto-electric-sizing units are used for more than 25 years now. These units use the optical properties of the fruit *viz.* reflection, transmittance or delayed emittance. With these sizers dropping impact is reduced. About 20-25 tonne fruit are sorted per hour in large and efficient packing houses. In opto-electrical sizers, light source and detector combination is used. The horizontal sizing is done by relating fruit size to a time length required for a fruit to continue intersecting a light beam which traverses conveyer belt (Iwamoto and Chuma, 1981).

It was reported electro–optical sorter for colour and size sorting of citrus fruits. This system includes 5 microprocessors to control feeding, size detection, colour detection, discharge and master control. Size grading is accomplished by number of LED light beams blocked by projected area of fruit, while colour grading is conducted by using dual wavelength IVRF quality index at 880 and 650 nm. Accuracy of the system is reported to be 91-92 per cent at 24-48 fruit/min speed.

The most important requirement of good sizer is its ability to adjust the sizes and flow of fruit even when in operation if required. After sizing, fruit of only required grade can be washed, waxed, dried and packed. The other grade can be waxed and packed later or small sizes can be sent for juice processing.

4.21.6 Packaging

The content of each package must be uniform and contain only oranges of the same origin, variety, quality, size, same degree of ripeness and development.

The use of any substance tending to modify the natural characteristics of the oranges, especially its taste or smell is prohibited.

The oranges shall be presented as follows:

i) Arranged in regular layers in packages. It is mandatory for "Extra Class" and optional for Class I and II.

ii) Not arranged in regular layers in packages or in bulk bins. This presentation is only allowed for Classes I and II.

10 per cent tolerance only applies to fruit whose diameter is not less than 53 mm.

Packing is the last operation on packing line and place packing (manually placing fruits in the containers and arranging in layers) is most common for oranges and mandarins. For hand packing fruits are collected in tubs, chutes, bins or in long sliding tables. Fruits are arranged in the box to sit with desired orientation. The padding or paper board separators or paper pulp tray can be used between fruit layers. Like sorting, packing is another operation where trained personnel are required since this is the operation which makes direct impact on buyers when box/container is opened in the market. The convenience of packer is of utmost importance in designing of packing table and sitting/standing facility for easy reach of fruit and container/tray. The position of the box should be as near as possible to the body of the packer. This allows packer to look at the bottom and place the fruit properly and conveniently. The filled boxes are lifted and put on the conveyers/rails traveling sideways. The empty boxes and other materials like trays and separators should be available to packer at a stretch of the arm or at the most with one or two step movement. Automatic place-packing units are also operational in many packing houses in the US and elsewhere. The carefully sized and sorted fruit of desired number are lifted and placed automatically in the box in specific orientation. Box is then inspected and closed. In orange packing machine used in some of the facilities of Sunkist growers, California, fruits are metered in to pattern formers where they are picked up by spring loaded suction cups. The fruit is then moved over the carton and placed automatically. Machines with capacity as high as 315 cartons of 18 kg capacity per hr (Johnson, 1991) are operational. The actual speed is up to 200 cartons per hr. The machine reduces labour requirement by 75 per cent.

Limes and Lemons are jumble-packed/volume-filled sometimes. Jumble pack can be automatic or semi-automatic. Fruits for local markets are usually jumble-packed where prices are to be kept low. The automatic volume–fill or jumble–pack machine fills the box with presized fruit to the desired weight and shakes it to immobilize the fruit. The vibration of box helps in settling the fruit below the desired top level of the box so that any fruit protruding above the top edge is not pressed while closing. The box is weighed on the scale manually and fruit is added/removed so that box is filled to desired weight. Then the box is closed and sealed automatically. The fully automatic large machines have facilities as high as 400-600 boxes per hour.

Whatever may be the method of packing, it is to be ensured that fruit are immobilized and fitting in to the container space without pressing.

4.21.7 Marking and Labeling

Labeling with individual lable/sticker on individual orange fruit is done just before packing. The singulated fruit in cups are synchronized to pass beneath the cassette of pregummed labels. Each fruit gets a lable with gentle pressure. The inks made of food–quality dyes are also developed to stamp on the fruit, however, its acceptance depends on rules and regulations of the country where fruit is to be marketed. Recently new laser technology is developed to tattoo each fruit with which product look-up (PLU) code, 'country of origin' labeling (COOL), grower lot number and any other requested security information can be marked on the fruit directly. This helps in tracking and tracing of fruit source and settling any problem. It is designed by Durand-Wayland Inc. USA (Anonymus,2005).Most grocers require PLU labeling as the number can be punched in to get the price.

Pregummed labels with all required information are also stuck to the package box. The marking stamps of the date of packing or any special treatment to fruit, and packers name and address are also used.

In the state of Florida, all citrus fruit sold or shipped for consumption in fresh form need to be graded in a registered packinghouse in the state according to standards established by the Department of Citrus, and the grade of such fruit shall be indicated/ stamped on the carton. Brands or trademarks need to be properly registered with the department to represent state or U.S. grades. In accordance with prescribed rules of Department of Citrus, licensed citrus fruit dealers in the state shall be entitled to register labels, brands, or trademarks for grade identification purposes. The department maintains a record of all labels, brands, or trademarks registered for grade identification purposes.

6.0 Conclusion

Grading is a value addition and results in better returns to growers as compared to sale of non-graded fruit. Grading also provides an authentic and scientific basis in promoting and managing the marketing system. It serves as a realistic and common basis for market intelligence and reporting. It facilitates e-trading and also settlement of quality disputes between buyers and sellers. Packaging reduces bruising, crushing and moisture loss, prevents re-contamination of the product with spoilage organisms, reduces pilferage and maintains a sanitary environment during marketing and thus improves sale of produce. Grading is done manually as well as mechanically. Electronic machine vision technology, high-speed x-ray scanners and Near–Infrared (NIR) technology is used for automatic grading on the basis of external and internal fruit quality. This ensures perfect fruit quality to the buyers. The Indian "Agmark" standards and the standards of the Codex Alimentarius Commission are similar to a large extent. For export, producers and exporters have to follow Codex standards apart from standards set by the country where fruit is being exported. There is a considerable development in the field of packaging. In consumer /retail packages, the concept of 'Smart packages' involves coating with certain patented chemical material that is sensitive to volatiles emanated by spoilage causing microbes. As the fruit starts decaying in package due to growth of spoilage organisms, the container colour changes due to coating which indicates that packed produce is no longer good for consumption. The intensity and hue of colour also indicates how long produce can be stored and safe for consumption. 'Active Packages' can be created by using oxygen scavengers, carbon dioxide absorbents / emitters, ethanol emitters and ethylene absorbents. The appropriate material is placed alongside the fresh fruit that modifies the atmosphere in the package and thereby contributes to the extension of shelf-life of the fresh fruit. New materials are being tried to develop eco-friendly sustainable material for packaging of produce. Continuous jute fibre as reinforcement in paper laminates to increase tensile properties of paper bags could be the substitute for plastic bags.

7.0 Important References

Anandswamy, B. and Venkatsubbaiah, G. 1976. Wooden and corrugated shipping containers for the export of Coorg oranges. *Indian Food Packer*, **30**(3): 44-49.

Anonymous. 1995. Central Institute on Postharvest Engineering and Technology (CIPHET), Ludhiana. *Annual Report for 1993-94 and 1994-95*, 31pp.

Anonymous. 2005. Tatooed fruit is on the way. *Processed Food industry*, New Delhi. **8**(10):40-41.

APEDA, 2007. Guidelines for export of Indian mangoes to USA. Agricultural and processed Food Products Export Development Authority, New Delhi. p.45.

Artés, F., Villaescusa, R. and Tudela, J.A. 2000.Modified atmosphere packaging of pomegranate. *J. Fd Sci.* **65**: 1112-1116.

Blandini, G., Levi, P., and Pappalardo, R. 1992. Problem analysis of an automatic citrus grading machine. *Proc. Intn. Soc. Citric.*, Italy pp. 1033-1035.

Chauhan, K.S., Sandooja, J.K., Sharma, R.K. and Singhrot, R.S. 1987. A note on assessment of certain prevailing practices for marketing of commercial fruits. *Haryana J. Hort. Sci.* **16:** 229-232.

Chen, S.M., Fon, D.S., Hong S.T, Wu, C J, Len, K.C, Tien, B.T. and Chang, W.H. 1992. Electro-optical citrus sorter. *Am. Soc. Agric. Engg. No.* 92-3520.19 pp.

Chopra, S.K. and Dhar, R.P. 1990. Packing and storage of citrus with special reference to Kinnow. *Proc. Seminar on Problems and Prospects of Kinnow cultivation*, PAU, Ludhiana, pp 205-209.

Codex Alimentarius Commission,2007. Codex standards for fresh fruits and vegetables, (First Edition), WHO and FAO of the United Nations, Rome. Italy, p. 185.

Dash, K.K., Goswami,T.K. and Deka, S.C. 2012. Modeling for the modified atmosphere packaging of sapota fruit. *Agric Engg Int.*,**14**: 246-255.

Agmark, 2012. Fruits and Vegetables Grading and Marking Rules. 2004, 2007. and 2012. Gazette of India, GSR, 220, and GSR 171.

Iwamoto, M. and Chuma, Y. 1981. Recent studies on development in automated citrus packinghouse facility in Japan. *Proc. Int. Soc. Citric*,**2**:831-834.

Jain, P.K. and Chauhan, K.S. 1994. Influence of packaging and modes of transportation on marketability of Kinnow. *Indian J. Hort,* **51:** 251-254.

Johnson, T.M. 1991. Citrus postharvest technology to control losses. *Proc Intn Citrus Symp, China,* pp. 704-708.

Ladaniya, M.S. 2001. Response of 'Mosambi' sweet orange (*Citrus sinensis*) to degreening, mechanical waxing, packaging and ambient storage conditions. *Indian J. Agric. Sci.,***71:** 234-239.

Ladaniya, M.S., Naqvi, S.A.M.H. and Dass, H.C. 1994. Packing line operations and storage of Nagpur mandarin. *Indian J. Hort.,***51**: 215-221.

Ladaniya, M.S. and Shyam Singh. 2000. Influence of ventilation and stacking pattern of corrugated fibreboard containers on forced-air precooling of 'Nagpur' mandarin.*J. Fd Sci. Technol.,***37**: 233-237.

Ladaniya, M.S., Sonkar R.K., Shaikh, A.J. and Varadarajan, P.V. 1999. Corrugated fibre board containers for packaging, transport and storage of 'Nagpur' mandarin in domestic market. *Indian Food Packer,***53**(4):5-15.

NHB. 2011. Indian Horticulture database 2011. National Horticulture Board, Gurgaon, India, p.281.

Pandiarajan, T, Pugaledhi, S, Devadas, C.T. and Gothandapani, L. 1993. Development and testing of lime fruit grader. *National Seminar on optimization of production and productivity of acid lime*. Horticulture College and Research Institute, Periyakulam, Tamil Nadu.

Porat, R., Weiss, B., Fuchs, Y., Sandman, A., Ward, G. and Kosto, I. 2008. Keeping quality of pomegranate fruit during prolonged storage and transport by map: new developments and commercial applications. *Acta Hort.*, **804**:115-120.

Shaikh, A.J, Ladaniya, M.S, Varadarajan, P.V. and Shyam Singh. 2003. Paper and corrugated boxes from cotton plant stalks for effective packaging of oranges. *J. Sci Industrial Res.,***62***:* 311-318.

Sukhvinder Pal Singh, 2003. Modified atmosphere packaging of papaya (*Carica papaya* L.) fruits for extension of storage life. M.Sc. Thesis. University of Agricultural Sciences, Bangalore.

Varshney, A. C., Mangaraj, S. and Birla, S. L. 2005. Grading, packaging and marketing of sapota in south Gujarat. *Agricultural Engineering Today,***29** (3-4): 40-47.

Verma, B.B. 2009. Continuous jute fibre reinforced laminated paper composite and reinforcement-fibre free paper laminate. *Bulletin of Material Science,* Indian Academy of Sciences, **32** (6): 589–595.

Chapter 8

Insect Disinfection and Quarantine

Abraham Verghese

Director, ICAR-National Bureau of Agriculturally Important Insects, Post Box No. 2491, H.A. Farm, Post Bellary Road, Bengaluru – 560 024, Karnataka, India

E-mail: avergis@iihr.ernet.in

M.A. Rashmi

Scientist, C/o Department of Entomology, University of Agricultural Sciences, GKVK, Bengaluru – 560 065, Karnataka, India

E-mail: rashmigowda.ento@gmail.com

1.0 Introduction

Fruit flies are a large and important group of insect pests that globally attack a wide range of fruits and vegetables, with significant economic implications for horticultural production and market access. Fruit flies do not recognise commercial/domestic or state, regional boundaries and jurisdictions. They are highly mobile, capable of flying short distances but can be transported to long distances through fruits. They also have the potential to infest a wide range of crops from mango to apple and tomato to grape. Also, unlike many other plant pests, the management of these pests cannot be conducted solely on farm and the loss of control on a particular plantation can have a wide implication for commercial regions and the wider community.

Most of the Asian Pacific regions have witnessed rapid growth in horticultural development. Changes in dietary habits owing to increasing incomes continue to accelerate demand for fruits and vegetables in the region. Among the important pests of horticultural crops, fruit flies (Tephritidae: Diptera) rank among the world's most serious pests. Although, many insect pests attack fruits and vegetables, none have garnered greater notoriety than Tephritid fruit flies which are recognized worldwide as the most important threat to the horticultural industry (Allwood and Drew, 1997; Barnes, 2004). Besides causing direct losses in yield and marketability, they are barriers to international trade of fresh fruits and vegetables.

Fruit flies may gain entry to a region where they are not yet present by several pathways. The risk of introduction through imports is generally considered low because imports of horticultural produce are regulated by strict quarantine requirements. The unauthorized movement of fruits and vegetables coming from untreated areas and carried by sea and airline passengers present the highest "risk pathway" for fruit fly introductions. The aim of this paper is to review the quarantine risks posed by major exotic fruit flies is to create awareness of the potential threats posed by these pests, and to highlight the management and phytosanitary issues.

2.0 Mango Exports

India is the largest producer of mangoes in the world and has a large opportunity to tap global trade. Beside this, relatively long period of availability (March to August), diversity of varieties, economic liberalization and priority to export of fresh products have opened up the possibility of boosting mango exports from India.

Both fresh mango and processed mango products are exported to different countries namely United Arab Emirates, Bangladesh, United Kingdom, Saudi Arabia, Kuwait, Qatar, Bahrain, Nepal, United States of America, Oman, Singapore etc and the foreign exchange earned from such exports amounts to about Rs. 2005 millions.

Mango production worldwide, but only 2 per cent of its production is exported. Other major mango producing countries, like Mexico, Brazil, and Pakistan, export much higher quantities of their production. The reason may be that small farmers own majority of India's mango plantations gardens and 72 per cent of the farms are less than three hectares. The other reason may also be that nearly 15 per cent of India's mango production is wasted due to lack of adequate infrastructure facilities. In addition, India's mango exports also suffer from other factors such as heterogeneous quality and phytosanitary issues due to lack of adequate postharvest operations. Indian farmers are not properly informed about internationally practiced postharvest operations for mangoes, which result in not only heavy postharvest wastage but also low level of price realization in international markets.

The scenario of export of mangoes may be altered by efficient orchard management, improvements in infrastructure, use of advanced technologies, and meeting the food quality standards such as ISO9001, HACCP, with India becoming the leading exporter.

2.1 Importance of Fruit Fly

No doubt, there is a huge market to capitalize its potential, exporters will have to comply with the quarantine requirements by adapting quality standards throughout value chain *i.e.*, preharvest and postharvest requirements like plant quarantine, preshipment and treatment, phytosanitary certification, presurveillance and monitoring, irradiation treatment, verification, and certification of vapor heat treatment, hot water dip treatment, and registration of packing house facilities. For the convenience of exporters, Govt. of India has instituted an elaborative mechanism, which includes all these operations. The early harvested crop of mangoes (April-May) in India is well suited to international demand, as there is less competition from other countries during that period.

Before the shipment the consignment are inspected by officers of the Directorate of Plant Protection, Quarantine and Storage to ensure compliance with phytosanitary requirements of importing country. The National Plant Quarantine Centre located at New Delhi, while there are four regional quarantine stations located at Amritsar (North), Chennai (South), Kolkata (East) and Mumbai (West). The Ministry of Agriculture undertakes regular pest survey and monitoring of mangoes, meant for export under National Fruit Fly Surveillance Programme. During 2011-12 the country exported mango is 63441.288 MT of mangoes to 41 countries of these, the 15 major countries importing mangores from India are listed in Table 1.

Table 1: Country-wise Export of Mango (Quantity in kg)

Country	*2009-10*	*2010-11*	*2011-12*
Bahrain	12,38,495	9,82,332	6,23,689
Bangladesh	3,35,49,895	2,30,49,685	2,75,99,483
Canada	2,55,193	4,26,682	4,05,759
France	2,07,365	1,14,884	1,16,864
Kuwait	8,04,151	5,77,044	7,31,243
Malaysia	4,15,309	3,97,821	3,53,413
Nepal	40,58,151	19,91,257	39,25,742
Netherland	94,353	22,426	2,56,851
Oman	2,69,500	1,36,756	1,54,774
Qatar	6,59,023	3,77,099	8,16,102
Saudi Arabia	31,47,133	15,82,378	23,88,634
Singapure	3,67,577	3,91,297	5,99,273
UAE	2,56,08,155	2,54,74,530	2,20,13,880
United Kingdom	29,58,655	27,21,906	25,32,422
United States	1,75,401	99,699	3,53,177

Source: DGCIS Annual Export.

Though India is the chief mango producer, Indian mangoes are grossly inadequate in the international market as export of fresh fruits is less than 5 per cent. Though cultivation of mango is very high, its consumption in local and foreign markets is comparatively less due to infestation by insects and poor shelf life in mangoes.

2.1.1 Major Insect Pests of Mango

a) The Oriental Fruit Fly, *Bactrocera dorsalis* (Hendel) (Diptera: Tephritidae)

It is a major pest of commercially important varieties of mangoes. It is distributed throughout India and undergoes hibernation during winter in north India; it is serious and active throughout the year in southern parts of India especially when the minimum temperature surges in summer. It is widespread in different parts of the world and is reported on more than 150 hosts. It was reported that the yield loss due to *B. dorsalis* (Figures 1) in India ranged from 1 to 31 per cent with a mean of 16 per cent.

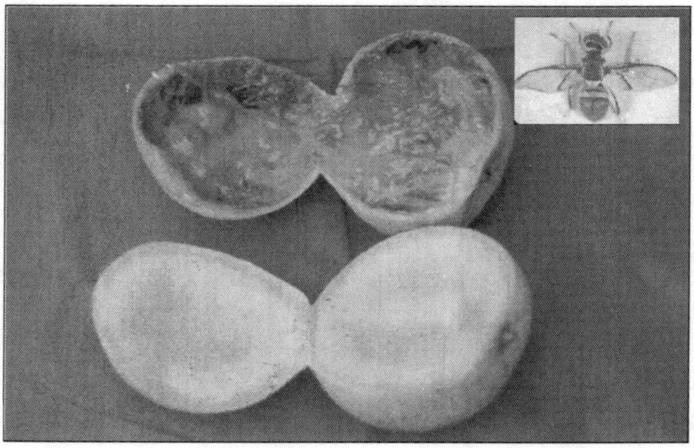

Figure 1: Fruit Fly Damage in mango by *Bactrocera dorsalis*

European and Mediterranean Plant Protection Organization (EPPO) lists *Bactrocera dorsalis* as an A1 quarantine pest (OEPP/EPPO, 1983) within Plant Health Committee of the broad category "non-European Tephritidae"; it is also of quarantine significance to Asia and Pacific Plant Protection Commission (APPPC). Southern Cone (COSAVE), Caribbean Plant Protection Commission (CPPC), Inter-African Phytosanitary Council (IAPSC), Cartagena Agreement Commission (JUNAC) and International Regional Organization for Agricultural Health (OIRSA). *Bactrocera dorsalis* is indigenous to Asia, but like other *Bactrocera* spp. is known by experience to have the potential to establish adventive populations in various other tropical areas. Besides *B. dorsalis*, *B. zonata* is quite important especially in the western and northern parts of India. (Appendix I mentions some important fruitfly of global significance).

b) Mango Stone Weevil, *Sternochetus mangiferae* (Fabricius)

Mango stone weevil is another important pest of mango with quarantine importance (Figures 2). It is a monophagous pest of with wide geographic distribution. It has univoltine life cycle, most of which, in India, is spent as resting adults from May/June to February/March. When mango fruits are developed to pea size (0.5-1cm diameter) the adults become active and begin feeding and mating. Oviposition on mango fruits begins when the fruits are lime size (2-4 cm diameter). The adult female makes a hollow, boat shaped depression, and spreads over the eggs. The larva hatches after an incubation period of 5-7 days and burrows through the flesh to reach the seed coat, leaving a visible track as a streak of water area leading to seed. Infected

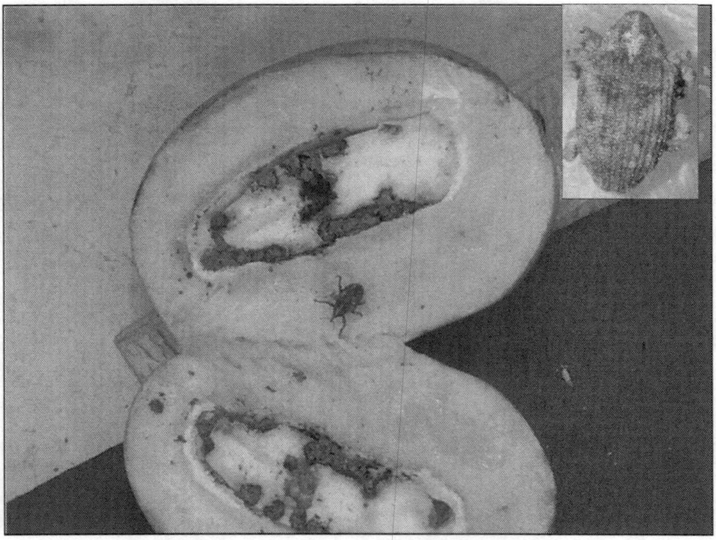

Figure 2: Damage by Mango Stone Weevil

fruits are difficult to detect since no damage is visible externally. Incisions made by the ovipositing females are small and heal as the fruit develops. Adults emerge by burrowing through the seed, pulp and rind, congregate in small numbers in cracks and crevices of mango stems, branches, leaf litter, clods, etc., and remain inactive until the next flowering season.

Direct loss due to mango stone weevil varies from 5 per cent to 80 per cent in different mango varieties. Infestation may cause indirect losses by stimulating the premature fall of developing fruits. If the weevil survives up to harvest, it affects the processing industry and export of fruit. Longer-range dispersal occurs largely through transport of fruit and seeds containing the larvae, pupae or adults. The pest has been intercepted in mango fruits and seeds in international trade.

Sternochetus mangiferae is a quarantine pest for the CPPC, IAPSC, NAPPO and OIRSA. It is not currently listed as a quarantine pest by the EPPO. However, as mango production increases in the south of the EPPO region, there is no doubt that *S. mangiferae* could establish and be of economic importance.

Fruits of mango containing larvae or pupae pose major risk of introducing *S. mangiferae* through the market chain. Although mango seeds with diapausing adults could also be a danger, such materials from countries where the pest occurs should be prohibited by mango-growing countries (EPPO).

2.1.2 Bottlenecks for Export

i) Complex supply chain

ii) Weather fluctuations, causing interrupted supply of fruits to the industry.

iii) Lack of supply of mangoes.

iv) Lack of forward and backward linkage between growers/suppliers and buyers/exporters

v) Inconsistent and inadequate supply of quality of mangoes.

vi) Disaggregate and multi-layered marketing channels, resulting in inefficiencies in procuring of raw material.

vii) Due to highly perishability and to pests and diseases in High postharvest losses

viii) Inadequate transport facilities high freight cost.

2.1.3 Pest Risk Analysis

Pest Risk Analysis (PRA) is the application of sound scientific principles to identify, evaluate and manage pest risks associated with trade and travel and is intimately associated with quarantine regulations that may hinder free trade and movement.

Pest Risk Analysis (PRA) is done to protect the country's agriculture from damages that can be caused by harmful (quarantine) pests which can be brought in along with imported commodities. Process consisting of three stages (i) initiation of the PRA through identification of a pest or pathway, or review or revision of an existing phytosanitary policy, (ii) pest risk assessment, and (iii) pest risk management. Risk communication is an integral component that occurs throughout each step.

PRA examines whether the criteria for the quarantine pest status of an animal or disease are justified. PRA examines the whole sequence of events that pests must survive if they are to become established in a new area and makes explicit and open the scientific judgements used. Such procedures are increasingly important as measure to liberalise world trade increase in importance. Agreement reached by signatories to the World Trade Organisation requires that "phytosanitary measure(s) are based on scientific principles and (are) not maintained without sufficient scientific evidence" (WTO, 1999).

A sound, rational and internationally justifiable and defendable quarantine policy must be based on pest risk analysis. PRA relies on a full understanding of the biology, ecology, behaviour and control of the potential pest species under consideration and places considerable technical demands in the gathering and interpretation of the data needed to support accurate assessments. As any assessment may be limited by the amount of information available about each species PRA should be used cautiously and critically, its very openness to examination and criticism being crucial to ensuring that assessments are soundly based.

The FAO define a quarantine pest as "a pest of potential economic importance to (an) area and not yet present there, or present but not widely distributed and being officially controlled". Clearly a major component of justifying quarantine regulations against a pest is the demonstration of clear economic importance. For this to occur a pest must become established and spread in an area and it is the evaluation of the risks associated with this potential establishment and spread that form the core of PRA.

The PRA results of risk of fruit flies in the Indian Ocean Region shows that exotic fruit species predominate and are more widely distributed than native species. These introduced species account for most of the damage caused by fruit flies to fruit crops and vegetables. Exotic fruit fly species have not only gained entry to the Region, but have been more successful in colonising and have led to the ecological suppression of native species such as *Ceratitis catoirii* and *Dacus demmerezi*.

Of fruit flies occurring outside the Indian context the fruit flies *Bactrocera, B. invadens* (Drew, Tsura and White), *B. papaya* (Drew and Hancock), *B. tryoni* (Froggatt), *B. neohumeralis* (Hardy), *B. philippinensis* (Drew and Hancock), *Ceratitis capitata, Dacus spp, Anastrepha sp, Rhagletus spp etc* are of most quarantine concern. Among these the genus with the greatest quarantine significance is *Bactrocera*. For example *B. dorsalis* is a global quarantine pest with complex of 52 sibling species, of which 40 are considered new species and at least 8 are of economic importance. This is a matter of great concern to fruit producing countries where these species occur, requiring them to re-evaluate and revise their control strategies. In countries where these species have not been recorded, quarantine measures have been considerably strengthened to minimise the risk of their entry, establishment and spread.

In Australia, a country where quarantine occupies top priority, 80 species of fruit flies, both endemic and exotic, have been identified as significant economic threats to Australian horticultural crops. The majority of these species are exotic to Australia, however many are found in close proximity throughout South-East Asia and the South Pacific. In addition, Australia has a number of endemic and established species. Two in particular, the Mediterranean fruit fly (*Ceratitis capitata*), which permanently inhabits the southern growing regions of Western Australia, and the Queensland fruit fly (*B. tryoni*), found in parts of the Northern Territory, Queensland, New South Wales and the eastern corner of Victoria. These species have a significant economic impact increasing costs of fruit production and market access.

In the 1970s the Asian species *Bactrocera carambolae* (Drew and Hancock) was found for the first time in the Western Hemisphere, in Surinam probably introduced through passenger luggage from Indonesia, and is the subject of an international control and eradication campaign. This in turn demonstrates that economically important American fruit fly species such as *Anastrepha ludens, A. fraterculus* and *A. obliqua* must also require consideration by the Quarantine Services of the Indian Region.

2.1.4 Sanitary and Phyto-sanitary Issues

Quarantine problems cannot be looked at in isolation. Regional and International collaborations are vital if progress is to be achieved in reducing risks of fruit fly introduction and spread. The introduction and spread of plants and pests of plants from one geographical area to another is an issue of worldwide concern, and is addressed at the international level by several agreements. The principal agreement aimed at preventing the spread and introduction of pests of plants and plant products is the International Plant Protection Convention (IPPC).

A quarantine pest is defined by the IPPC as "a pest of potential economic importance to the area endangered thereby and not yet present there, or present but not widely distributed and being officially controlled" (International Plant Protection Convention, 1997). In other words, it is any organism that is injurious or potentially injurious, directly or indirectly, to plants or plant products or by-products of plants and includes bacteria, fungi, insects, mites, molluscs, nematodes, other plants and viruses, not present in a specified area at risk or, if present, being controlled by an NPPO (National Plant Protection Organization).

The invasion potential of fruit fly is determined by its likelihood to be transported and carried from one place to the other, and by its ability to get established in the new environment. Once a fruit fly reaches the new environment it requires to have the plasticity to adapt to the new physical conditions, and be able to reach its hosts to reproduce.

Phytosanitary measures to enhance mango exports in terms of pest free areas (PFA) have been undertaken at IIHR, Bangalore for *Bactrocera caryeae* (Kapoor) and commodity based PFA for *Bactrocera spp*. including *B. dorsalis.* As part of the trade efforts there is a need to bring simple identification keys for quarantine pests. At IIHR, biosystematics of common fruit flies has been taken up and a key has already been published (Madhura and Verghese, 2004). Further acceptable postharvest treatments are essential to export and import fresh mango fruits.

2.2 Requirements for Exporting Mangoes to United States

An APHIS preclearance officer, in India, first inspects mangoes for presence of pests so that plant pests do not enter the United States. The fruits are then treated with specified doses of irradiation prior to export. After passing preclearing requires, if, live pests are found during the re-clearance inspection, the shipment is rejected and is not eligible for export to the United States. Mangoes must be packed in insect-proof boxes and safeguarded after treatment to prevent re-infestation. It must also be accompanied by a phytosanitary certificate issued by the national plant protection organization (NPPO) of India with additional declarations certifying that the treatment and inspection of the mangoes are made in accordance with the regulations.

2.3 Strategy for Export

The strategies for boosting up of mango exports suggested by the Expert Group of the Planning Commission and APEDA are as under:-

i) Trans-shipment of mangoes by sea to make them competitive in price sensitive markets of Europe and Middle East.

ii) Promotion of non-traditional varieties like Chausa, Safeda and Badami.

iii) Augmentation of supply during seasonal gaps. (February/March in Middle East and March/April and July/August in Europe).

iv) Development of postharvest management facilities/practices (DMI, 2013).

3.0 Globally Accepted PHM Practices

The presence of arthropod pests in or on horticultural commodities has caused major disruptions in the shelf-life, processing, shipment of fruits and vegetables and exports. Postharvest disinfestation treatments can be used to control the presence of insects, so that the risk of rejection at importing countries as a result of insect contamination can thereby be reduced. These treatments should control the pest species without damaging the crop. However, there is normally only limited opportunity between combating the insects and damaging the fruit.

The challenge of research on postharvest quarantine treatments is that the treatment must not only be effective against virtually all of the pest individuals present on the fruit, but also minimize damage to the commodity being treated. Additionally, the treatment has to be commercially viable *i.e.* it must be practical, economical and amenable to regulatory approvals at the importing destinations. The main techniques available are fumigation, usually with methyl bromide, heat treatment, either with hot vapour or hot water, cold treatments, insecticidal dipping and irradiation. Current technology has led to the development of numerous means to control (chemical, mechanical or procedural) these pests, at preharvest. For most part, postharvest pest control is focused on trade and exports. The accidental introduction or fear of introduction of a pest into a place where it is not known to be present has resulted in the establishment of quarantine restrictions and export treatment requirements.

Postharvest heat treatments are being used for disinfestations and disinfection of an increasing variety of flowers, fruits, vegetables and flower crops. Postharvest management determines food quality and safety, competitiveness in the market, and the profits earned by producers. Postharvest treatments can include conventional chemicals, physical, irradiation and organic treatments.

3.1 Physical Methods

Physical methods of quarantine and phytosanitation treatments include mechanical, ultrasound, vacuum and temperature extremes. Fumigants are the most prevalent, but are losing support because of health, environmental and safety concerns. Quarantine systems using fumigants such as ethylene di-bromide (EDB) and methyl bromide (MB) are not acceptable anymore by almost all importing countries.

Vacuum treatments, applied primarily to durable commodities, are gaining interest due to the development of flexible, low- cost treatment containers. Vacuum treatments appear to cause mortality in insects by creating a low-oxygen atmosphere and in low-moisture environments and may cause rapid dehydration. For entry into USA, the mangoes need to be treated in India with irradiation by receiving a minimum absorbed dosage of 400 Grays. Each consignment of mangoes must be accompanied by a phytosanitary certificate issued by the National Plant Protection Organization (NPPO) of India.

3.2 Ionising Radiation

The development of non-chemical techniques to maintain postharvest quality of fruits and vegetables is increasingly emphasized in many researches. Perhaps the most successful and widely accepted nonchemical quarantine treatment is the use of ionising radiation. Although irradiation is considered a physical treatment, it is not commonly considered an 'organically compliant' measure, even though it does not render the commodity or affected arthropods radioactive nor does it leave any detectable residues deemed to be harmful to human or animal health.

3.3 Chemical Treatment

Chemical treatments such as fumigants, pesticide dips and detergents are considered conventional and undesirable since they do not fit into the organic health and environmental standards of many countries. Among the many chemical fumigants identified for controlling postharvest arthropods are phosphine and sulphuryl fluoride (SF). But these are not advocated for fruit flies in mango.

4.0 Postharvest Treatments

4.1 Cold Water Treatment

Cold temperature treatments particularly cold storage have practical application for many commercial operations such as holding commodities before marketing, as well as pest control. Thermal treatments must be precise because of the narrow margin between efficacy and commodity tolerance to the temperature, particularly in fresh fruit and vegetables.

The fruits are subjected to cold storage quarantine treatment when shipped to areas outside of the known range and where the fly could survive. Rapid cooling in cold water increases the mortality of Caribbean fruit fly *Anastrepha suspensa* (Loew), larvae in carambolas.

In India cold treatment experiments at 8 °C exposure for a week, showed zero per cent infestation in Banganapalli and Totapuri mango, while Alphonso showed 10 per cent infestation. This has treatment the potential to act as ovicide. Varieties levels were 25, 40 and 40 per cent in the Banganapalli, Alphonso and Totapuri respectively.

4.2 Hot Water Treatment

Hot water immersion quarantine treatments are relatively easy to engineer. Its use as a quarantine system was intensified after the banning of ethylene di bromide. In Mexico, this quarantine system started in 1988 for the control of fruit fly (*Anastrepha ludens* and *A. obliqua*), mango fruits exported to USA and Japan, and consists of immersion of fruit in water at 46.1 °C (115 °F) for 65, 75 or 90 minutes, depending on the weight of the fruit. This system is now widely used in different mango growing regions in Mexico, Central and South America, and the West Indies. This system did not reduce the mango exports as anticipated and rather increased it. Other countries outside of America have also started experimenting and using hot water quarantine systems. It has been found to be effective in the disinfestation of several insects, other than the genus *Anastrepha*. Hot water treatments can damage the fruit. In the first few seasons the treatment caused major fruit injury in Mexico. However, now the treatment is used much more properly and therefore the damage has been controlled. After quarantine treatment, fruit must be introduced to an isolated, insect-proof area to prevent re-infestation. From here on (packaging, transport, etc) and until the fruit is received at the importing end, secure measures are implemented to prevent reinfestation. After heat treatment, fruit is usually cooled with water at ambient temperature and/or forced air cooling. In Mexico, fruit is first cooled in ambient water (at temperatures of < 21 °C), packed, and then cooled with forced-air cooling.

For exports of mango, the treatments for the Oriental fruitfly, *Bactrocera dorsalis* should ensure 99.99683 per cent control (By 'probit 9' standards). Heat treatment of infested fruits is known to ensure this standard. It has been reported that the eggs showed total mortality at temperature- time regimes of 45 °C -60 min and 48 °C - 45 min and 60 min. The first instar maggot was completely affected at 48 °C -60 min while the second instar maggot succumbed to 100 per cent mortality at 45 °C and 48 °C for 45 °C min and 60 min. Further the temperature regime 47 °C ± 1 °C for 75-90 min was validated in different varieties namely Alphonso, Banganapally, Neelam and Totapuri. This is the most cost effective method to disinfest exportable fresh, provided it is done within 24hrs after harvested and desapping.

Hot water plant for fruit fly is yet to be commercialized in India. The approximate cost of a one tonne capacity thermal tank will range vary depending on type of material used (zinc sheet, steel, fibre glass etc.,) and source of energy (electricity/ solar). To treat one kg of fruit the approximate cost, including labour has been reported to be less than Rs.2.0.

4.3 Vapour Heat Treatment

Vapour heat, the oldest of the quarantine heat treatment, consists of heating the host fruit by moving hot air saturated with water vapour over the fruit surface. When the temperature of the mango fruit is at or below the dew point, condensation of atmospheric moisture occurs on the surface of the fruit. In this way, fruits are heated by conductive energy transfer. The heat from the fruit surface is transferred towards the fruit centre. Commercial facilities operate in Philippines, Thailand, the United States and Australia. The VHT disinfestations protocol accepted for mango access to the high-value markets in Japan include

i) 46 ° C fruit core temperature held for 10 min for Philippine 'Carabao' mango

ii) 46.5 ° C fruit core temperature held for 30 min for Taiwanese 'Irwin' and 'Haden' mangoes

iii) 46.5 ° C fruit core temperature held for 10 min for Tailand 'Nang Klang Wun' mangoes

iv) 47 ° C fruit core temperature held for 10 min for 'Nam Doc Mai', 'Pimsen Dang' and 'Rad' mangoes.

Vapour heat treatment requires that mango be treated with vapor heat until surface temperature is 46 °C and seed temperature reaches 46.5 °C and held that way for 10 minutes. A vapour heat treatment has also been approved for the control of the Queensland fruit fly in Kensington mango exported from Australia to Japan. The treatment consists in raising the 46.5 °C core temperature held at 10 minutes. The USA also approved a quarantine vapour heat treatment for the control of Mexican fruit fly (*Anastrepha ludens*) and other *Anastrepha* species in Manila mango, and for mango from Taiwan infested with the oriental fruit fly.

Vapour heat treatment has also been approved for cv Alphonso, for export to Japan from India. Through efforts of ICAR and APEDA Vapour Heat Treatment Plants have been set up at important mango belts to ensure export to Japan. The current cost of the equipment is Rs.2-4 crores, and the equipment is supplied by Japan.

4.4 Forced Hot-Air Heating Treatment (FHAT)

Forced hot-air, also known as high-temperature forced air, is a modification of the vapour heat treatment developed by Armstrong *et al.* (1989) in to kill Mediterranean fruit fly, melon fly and oriental fruit fly eggs and larvae in papaya. It is essentially the same as vapour heat except that the fruit surfaces are dry during forced hot-air treatment. Improvements in temperature and moisture monitoring and air delivery have advanced forced hot-air treatments, leading forced hot-air treatments to be developed for commodities previously treated with vapour heat and also being developed for new commodities. Forced hot-air treatment appears to be as effective in controlling internal pests as vapour heat, and provides better fruit quality, becoming the treatment of choice for many fruit previously treated with vapour heat. The fruit skin temperature remains cooler during forced hot air treatments than during vapor heat treatment while the tissue just below the skin heats to lethal temperatures because of the occurrence of evaporative cooling on the fruit surface during forced hot air treatment at lower relative humidity. The hot air is usually forced over the surface of the fruit which slowly heat the pulp. When air at 50 °C is forced over mango surface all stages of mango fruit fly, West Indian fruit fly, and *Anastrepha serpentina* are killed when the seed surface temperature reaches 48 °C. Mango fruit weighing over 700 g cannot be treated with this system.

However, commercial usage of FHAT (forced hot air treatment), currently occurs only for Papaya grown in Hawaii and exported to mainland United sates, and in the Cook Islands where papaya is treated before being exported to New Zealand. This is not a treatment available in India, hence economics of this are not available.

4.5 Irradiation

Irradiation has been amply shown to have potential as a quarantine treatment. The reason it is not currently being used is frequently attributed to public distrust. Although application of ionizing irradiation as a quarantine treatment against tephritid fruit flies has been modest, the research record is extensive. Koidsumi in 1930 seems to have been the first to study ionizing radiation as a quarantine treatment over 70 yr ago. The measure of efficacy of an ionizing radiation disinfestations treatment against tephritids is normally prevention of adult emergence from eggs and larvae that are irradiated in fruits.

Efficacy of quarantine treatments against fruit flies usually must be confirmed by treating large numbers of insects of the most tolerant stage present in the commodity with no survivors. In the United States, this requirement has historically been interpreted to mean that at least 93,600 fruit flies of the most radio tolerant stage should be treated with no survivors to achieve the level of control of LD 99.9968 (referred to as probit 9) at the 95 per cent confidence interval.

Irradiation of fruit flies for development of quarantine treatments has been done in three ways: (i) *In vitro* with or without a semi-artificial rearing medium, (ii) rearing on one medium followed by insertion into a fruit for irradiation, and (iii) rearing and irradiation in the fruit. Fruit flies irradiated using the first two techniques have often been reported to be more susceptible to irradiation than those reared and irradiated inside fruits; the latter technique, being the closest to the natural situation, is more appropriate for irradiation quarantine treatment research.

Research using gamma radiation for the disinfestation of oriental fruit fly, *B. dorsalis,* in mangoes and its effect on the overall quality and acceptability of the treated fruit undertaken in the Philippines showed that mature larvae of the fruit fly were the most tolerant to irradiation, with the young eggs being the most sensitive. A minimum dose of 100 GY was required to prevent the emergence of adult fruit flies and to maintain quarantine security against the possibility of introducing this pest into the importing country. In contrast to vapour heat treatment, no internal breakdown was observed, even if irradiated at 350 GY. Therefore, use of irradiation at a minimum dose of 100 GY as a quarantine treatment for the oriental fruit fly in mature green mango fruits has been recommended. In order to reduce fruit damage and speed treatment quarantine treatment doses should be as low as is possible, provided that sufficient substantiating data support efficacy.

In Hawaii, packaged mangoes treated with cobalt-60 gamma irradiation and forced that minimum doses of 20.6 and 32.9 krad killed insects of all stages within the fruits which the few surviving adults were sterile and short-lived. In South Africa, irradiation of ripe, marketable fruit protected it from damage and prevented adult emergence. The most effective dosages ranged from 0.5 to 0.85 kGy.

There has been some variation in the precise definition of efficacy of irradiation as a quarantine treatment against fruit flies. Efficacy with irradiation has been defined as prevention of emergence of adult flies capable of flight and prevention of flies capable of reproducing (APHIS, 1989). These seemingly minor discrepancies may result in different doses required. Of the above three criteria, prevention of reproduction may be achieved with lower doses than prevention of adult emergence with prevention of adults capable of flight requiring a dose between those two. Currently APHIS bases efficacy on prevention of adults capable of flight in order to prevent any flies surviving a quarantine treatment from reaching detection traps and needlessly triggering eradication programs (Table 2). Data on cost benefit ratio are not available, though if trade to US and Australia, the benefit accrued will be high.

4.6 Microwave or Radio Frequency Treatment

The microwave region of the electromagnetic spectrum is from 1 to 100 GHz, between infrared and FM radio, and is close to the radio frequency range. Radio frequency (RF) waves are at the lower frequency range of the electromagnetic spectrum, with longer wavelengths. Accepted frequencies for industrial purposes are 13.56, 27.12, and 40.68 MHz. RF energy generates internal heat by agitating molecules in the fruit with a very rapid change in charge within the electrical field. The advantages of RF heating are that it is very fast, can penetrate deep, and can sometimes heat insects more than fruit.

The time required to increase fruit centre temperature to 47 °C in vapour heat treatments can be 45 minutes or longer, resulting in fruit susceptibility to heat damage.

Table 2: Minimum Absorbed Doses Suggested by USDA-APHIS for Quarantine Security of Fruit Flies in different Crops (APHIS, 1996)

Species	*Common Name*	*Dose (Gy)*	*Geographical Distribution*	*Crop*
Anastrepha ludens (Loew)	Mexican fruit fly	150	Extreme southern Texas to Guatemala	Grape fruit, Orange, Pear, Peach
A. obliqua (Mac- quart)	West Indian fruit	150	Caribbean islands, Mexico to Brazil	Cashew, Pawpaw, carambola, Sour orange, Grapefruit, Arabica coffee, Mango
A. serpentina (Wiedemann)	Zapote fruit fly	150	Mexico to Argentina	Calamondin, Pond-apple, Sweet orange, Peach, Gauva
A. suspensa (Loew)	Caribbean fruit fly	150	Florida, Caribbean islands	Gauva, Peach, Roseapple
Bactrocera cucurbitae (Coquillett)	Melon fly	210	Asia, parts of eastern Africa, Hawaii	Cucurbits-Cucumber, Gourds, Melons
B. dorsalis (Hendel)	Oriental fruit fly	250	India to southern China, Hawaii, N. Mariana Islands	Mango, Gauva, Papaya, citrus, fig, apricot, Annona
B. jarbisi (Tyron)		150	Northern Australia	Mango
B. latifrons (Hendel)	Malaysian fruit fly	150	India to China, Laos, and Singapore; Hawaii	Capsicum, tomato, Papaya
B. tryoni (Froggatt)	Queensland fruit fly	150	Australia, New Guinea, New Caledonia, Austral Islands, Society Isflylands	Annona, Avocado, Citrus, Mango
Ceratitis capitata (Wiedemann)	Mediterranean fruit fly	225	Mediterranean, Africa, Central and South America, Middle East, Hawaii, Western Australia, N. Mariana Islands	Ackee, Orange, Star apple, Coffee, Fig, Gauva, Peach, Apricot

Rapid heating by microwave or radio frequency energy can reduce the potential for fruit damage. The concept of high-temperature-short-time treatment is possible to shorten the treatment time while retaining fruit fly control at the Probit 9 (quarantine security) level. This is not a treatment available in India; hence economics of this are not available.

5.0 Integrated Pest Management

5.1 Fruit fly

The Indian Institute of Horticultural Research (IIHR), Bangalore, has worked out cost-effective and residue-free control strategies for the integrated pest management (IPM) of fruit flies and the technical know-how is as follows:

i) Fruit fly infestation begins 45- 60 days prior to harvest. Initial breeding of the fruit flies takes place in fallen fruits. So, collect and destroy [by deep burying (Figure 3) (at least 2 feet) or burning] all fallen fruits at weekly interval, from two months prior to harvest.

ii) Place fruit fly traps, developed with the know-how of IIHR @ 15-20 per hectare. Before placing the traps in the field add 5 drops of melathion or dichlorvos on the plywood pieces impregnated with pheromone. Traps should be fastened well on lower branches, between 2 meter height, at least 30 days prior to harvest or earlier.

iii) Traps should be kept in shade, to enhance their life.

iv) After 12- 15 days of trap placement, apply again 5 drops of melathion or dichlorvos on the plywood using ink filler to reactivate the trap (Figure 4).

v) Replace the used plywood with fresh plywood lure every 3-4 weeks, if harvest is prolonged, or if there are late varieties like Neelum still in fruiting. (If trap is full, count and empty the dead flies).

vi) If trap monitoring shows more than 5 flies/day, there is need to give three bait splashes on trunk (Figure 5), starting 3 weeks prior to harvest

vii) Bag with cloth lined covers 3 weeks prior to harvest to prevent infestation

viii) Postharvest disinfestation (within 24hr of harvest/after desapping) in water at 47 °C±1 for 75 – 90 min.

5.2 Stone Weevil

i) The adults spend nearly 6-8 months in hiding in junctions of trunk and primary branches between July and February where mud and leaf litter gather and decompose. A good cultural method to control the weevils is to clean these junctions with a hard broom any time after July/August. This may be repeated in December/January.

ii) When any chemical sprays like imidacloprid is given for hopper (*Idioscopus* or *Amritodus)* control, farmers should ensure good coverage on trunk, branches and canopy. Not only this would control the hoppers (even those hidings in the barks) but also kill resting adult stone weevils (Note: Adult weevils rest on mango trunk junctions approximately from July to February).

Figure 3: Burying of Infested Fruits

Figure 4: Fruit Fly Trap

Figure 5: Bait Splash on Trunk

Figure 6: Field Sanitation

iii) When 50 per cent of the fruits have attained lime size (2-4cm diameter) give a spray of acephate 1.5g/l followed after two weeks by decamethrin 1ml/l.

(Note: Wherever fruits are intended for export, avoid acephate and instead give two sprays of decamethrin) (Verghese and Shivananda, 2010)

6.0 Conclusion

Complex supply chain is one of the main barriers to development of the Indian Mango industry. In markets a number of small producers, commission agents are operating. Therefore, for a bulk purchase of lot, these numbers of functionaries are to be dealt, causing inconvenience and more postharvest losses.

Mango fruit flies and Stone weevil are the major quarantine pests from India. Of these two, fruit flies are serious pests of agriculture throughout the world and represent a threat to the horticulture and ecology of India. The presence of arthropod pests in or on horticultural commodities has caused major disruptions in the shelf-life, processing, shipment of fruits and vegetables and exports. The fruit is cultivated in the largest area *i.e.* 2,312 thousand ha and the production is around 15.03 million tons, contributing 40.48 per cent of the total world production of mango. More than thousand varieties of Mangoes are grown in India. However, only about 30 varieties are grown on commercial scale in different states. During 2011-12, export of Mangoes was 63441 tonnes. Bangladesh, UAE, Saudi Arabia and Nepal were the important countries where Mangoes were exported.

Postharvest disinfestation treatments can be used to control the presence of insects, so that the risk of rejection at importing countries as a result of insect contamination can thereby be reduced.

As India imports apple from Chile, US, Australia, South America, New Zealand etc., inadvertent introduction of fruit flies and other pests should be expected.

The major potential invasives could be *Ceratitis capitata, B. aquilonis, B. jarvisi, B. neohumeralis, B. papyae, B. psidii, B. tryoni* being threat from Australia and New Zealand, *Anastrepha* sp. from South America, *Rhagoletis*, from US.

Its important that India maintain high vigilance of the imported fruits and they constantly ensure strict compliance of quarantine and comprehensive postharvest control programs by the exporting countries.

7.0 Important References

Allwood, A. J. and Drew, R. A. I. 1997. *Fruit fly management in the Pacific*. ACIAR Proceedings No 76. 267pp.

Amstrong, J.W. 1994. Heat and cold treatments. **In**: Paull, R.E. and Amstrong, J.W (eds) *Insect pest and horticultural products*: treatments and responses CAB international, Wallingford. UK pp.103-119.

Animal and Plant Health Inspection Service. 1989. Section III,9 and section VI-T106. Plant protection and quarantine manual. US Dept. Agric., Wash. DC.

Animal and Plant Health Inspection Service. 1996. Schedules for propagative plant material T200d. Dipping. PQQ Treatment Manual. US Dept. Agric., Wash., DC.

Anon, 2010. Second Annual Report of DBT Scheme, "*Prevention of Postharvest Spoilage Due To Fruit Flies In Mango*".

Anonymous, 2006, Food irradiation update. Food Technology. **60** (10): 73–75.

Armstrong, J.W., Hu, B.K.S., Brown, S.A., 1996. Single-temperature forced hot-air quarantine treatment to control fruit flies (Diptera: Tephritidae) in papaya. *J. Econ. Entomol.* **88**:678–682.

Armstrong, J.W., J.D. Hansen, B.K.S. Hu, and Brown, S.A. 1989. High-temperature forced-air quarantine treatment for papayas infested with tephritid fruit flies (Dipterna: Tephritidae). *J. Econ. Entomol.* **82**:1667-1674.

Balock, J.W and Kozuma, T.T. 1964. Notes on the biology and economic importance of the mango weevil, *Sternochetus mangiferae* (Fabricius), in Hawaii (Coleoptera: Curculionidae). *Proceedings of the Hawaiian Entomological Society*, **18**, 353-364.

Barnes, B. N., 2004. **In:** Fruit Flies of Economic Importance. *Proc. 6th Intl. Symp.* Isteg Scientific Publications, Irene South Africa.

CFFP. 1999. Carambola Fruit Fly Programme. (http://www.carambolafly.com)

Drew, R. A. I, and Hancock, D. L., 1994. The *Bactrocera dorsalis* complex of fruit flies (Diptera: Tephritidae: Dacinae) in Asia. *Bull. Entomol. Res. Supple.* 2.

Ekesi, S. and Billah, M. K., 2007. A field guide to the management of economically important tephritid fruit flies in Africa. ICIPE, Nairobi, Kenya.

FAO, 1999. *Guidelines for Pest Risk Analysis*. Food and Agriculture Organisation. (http://www.fao.org/docrep/x5585e/x5585e0g.htm

FAO, 2002. International Standards for Phytosanitary Measures. Glossary of Phytosanitary Terms, Publication No. 5. UN-FAO-IPPC, Italy.

FAO, 2007. FAOSTAT. Food and Agriculture Organization of the United Nations. FAO Statistics Division, Rome, Italy. http://faostat.fao.org/(accessed 18 May 2007).

Fields, P. G and White, N. D. G. 2002. Alternatives to methyl bromide treatments for stored-product and quarantine insects. *Annual Review of Entomology*. **47**: 331–359.

Follett, P. A. 2009, Generic radiation quarantine treatments: The next steps. *J. Econ. Entomol.* **102** (4): 1399–1406.

Hansen, J.D. and Johnson, J.A. 2007. Introduction. **In**: Tang, J., E. Mitcham, S. Wang and S. Lurie, Eds., *Heat Treatments for Postharvest Pest Control: Theory and Practice*, CAB International, Wallingford, Cambridge.

Jordan, R. A. 1993. The disinfestation heat treatment process. Plant quarantine in Asia and the Pacific. A Report of an Asian Productivity Organization Study Meeting, Taipei, Taiwan, 17-26 March 1992. Asian Productivity Organization, Tokyo, pp. 53 - 68.

Koidsumi, K. 1930. Quantitative studies on the lethal action of x-rays upon certain insects. *J. Soc. Trop. Agric. (Jpn.)*. **2**: 243-263.

Krista, C., Robert, S. L. and Mangan. 2000. Postharvest disinfestation heat treatments: response of fruit and fruit fly larvae to different heating media, *Postharvest Biology and Technology*. **21**: 51–60.

Madura, H.S. and Verghese, A. 2004. A guide to identification of some common fruit flies (*Bactrocera* spp.) (Diptera: Tephritidae: Dacinae). Pest management in horticultural systems, **10**:1-10.

Manoto, E. C., Resilva, S. S., Rosario, S. E. D., Casubha, L. C., Lizada, C. C., Esguerra, E. B., Brena, S. R. and Fuentes. R. A. 1992. Effects of gamma radiation on the insect mortality and fruit quality of Philippine ÔCarabaoÕ mangoes, pp. 91-116. **In**: *Use of irradiation as a quarantine treatment of food and agricultural commodities*. International Atomic Energy Agency, Vienna, Austria.

Navarro, S. and Calderon, M. 1979, Mode of action of low atmospheric pressures on *Ephestia cautella* (Wlk.) pupae. *Experientia*. **35**: 620- 621.

Navarro, S., Donahaye, J. E., Dias, R., Azrieli, A., Rindnere, M., Phillips, T., Noyes, R., Villers, P., Detrain, T., Truby, R. and Rodrquez, R. 2001. Application of vacuum in a transportable system for insect control in: Donahaye, E. J., Navarro, S and Leesch, J.G (eds) proceedings of international conference of controlled atmosphere and fumigation in stored products, Fresno, California, 29 october-3 nov 2000. Executive printing services, Clovis, California, pp. 308-315.

OEPP/EPPO (1983) Data sheets on quarantine organisms No. 41, Trypetidae (non-European). *Bulletin OEPP/EPPO Bulletin* **13** (1).

Schneider, S. M, Rosskopf, E. N, Leesch, J. G, Chellemi, D. O, Bull, C. T. and Mazzola, M. 2003. United States Department of Agriculture – Agricultural research service research on alternatives to methyl bromide: preplant and postharvest. Pest Management Science: **59** (6–7): 814–826.

Shellie, K. C. and Mangan, R. L. 2000. Postharvest disinfestation heat treatments: response of fruit and fruit fly larvae to different heating media, *Postharvest Biology and Technology*. **21**: 51–60.

Stonehouse, J. M., Verghese, A., Mumford, J. D., Thomas, J., Jiji, T., Faleiro, R., Patel, Z. P., Jhala, R. C., Patel, Shukla, R. K. R. P., Satpathy, S., Singh, H.S., Singh, A. and Sardana. H.R. 2005. Research conclusions and recommendations for the on–farm IPM of Tephritid fruit flies in India, *Pest Management in Horticultural Ecosystems*, **11**(2): 172-180.

Tang, J., J.N. Ikediala, S. Wang, J.D. Hansen, and R.P. Cavalieri. 2000. High-temperature-short-time thermal quarantine methods. *Postharvest Biol. Technol*. **21**:129-145.

U.S. Department of Agriculture [USDA]. 2005. Animal Plant Health Inspection Service (APHIS), Plant Protection and Quarantine (PPQ) Treatment Manual: regulating the importation of fresh fruits and vegetables, Riverdale, Maryland,306pp.

Uma Maheshewari and Purusotham, U.T., 1999. A simple method to diagnose nut weevil incidence in mango. *Insect Environ*. **5** (3),101

Varith, J., W. Sirikajornjaru, and R. Kiatsiriroat. 2006. Quarantine treatment on mango using microwave-vapor combined process. *Proceedings of the American Society* of Agricultural and Biological Engineers Annual International Meeting, Portland, Oregon, 9–12, July, 2006.

Verghese, A. and Devi, S. 1998. Relation between traps catch of *Bactrocera dorsalis* Hendel and abiotic factors. Pp. 15-18. **In**: *Proc. First Nat. Sym.* Pest Management in horticultural crops: Environmental Implications and Thrusts (Eds., Reddy, P.P., Krishna kumar, N.K. and Verghese, A). Association for advancement of pest management in horticultural ecosystem, Bangalore.

Verghese, A., 1998. Fallen fruit sampling helps timing of first spray for control of mango stone weevil, *Sternochetus mangiferae* Fab. (Coleoptera: Curculionidae). *Insect Environ*. **3** (4), 91-92.

Verghese, A., 2000. Status and management of the mango stone weevil *Sternochetus mangiferae* Fab. (Coleoptera: Curculionidae) in India. *Pest Manage. Hortic Ecosys.* **6**(1), 15-21.

Verghese, A., Madhura, H. S., Jayanthi, P. D. K and Stonehouse, J. M. 2002. Fruit flies of economic significance in India, with special reference to *Bactrocera dorsalis* (Hendel). *Proceedings of 6th International Fruit fly Symposium* held between 6 – 10 May 2002, Stellenbosch, South Africa. pp. 317 – 324.

Verghese, A., Madhura, H.S., Jayanthi, P. D. K, Stonehouse, J. M., 2003. Fruit flies of economic significance in India with special reference to *Bactrocera dorsalis* Hendel. In: Barnes, B., Addison, M. (Eds.), Proceedings of the Sixth International Symposium on Fruit Flies of Economic Importance, 6–10 May 2002, Stellenbosch, South Africa, in press.

Verghese, A., Sreedevi, K., Sudhadevi, D. K., Nagaraju, B. R. and Jayanthi, M., 2006, Pest Free Areas (PFA) of *Bactrocera caryeae* (Kapoor) in Major Mango Belts of South India *Pest Management in Horticultural Ecosystems*, **12**(2): 75-84

Verghese, A., Shivananda, T.N. and Hegde, M.R., 2010, Integrated pest management of fruit fly in mango. Hand bill.

Verghese, A. and Shivananda, T.N., 2010, Integrated pest management of mango stone weevil in mango. Hand bill.

Verghese, A., Nagaraju. D. K. and Sreedevi, K, 2011. Hot water as an effective postharvest disinfection for the oriental fruit fly, *Bactrocera dorsalis* (Hendel) on mango. *Pest management in horticultural ecosystems*, **17**(2): 63-68

Vijaysegaran, S., Lam, P. F., Yon, R. M., Karim, M. N., Ramli, A. H. and Yusof, N. 1992. Gamma irradiation as a quarantine treatment for carambola, papaya and mango, pp. 53-76. **In**: Use of irradiation as a quarantine treatment of food and agricultural commodities. International Atomic Energy Agency, Vienna, Austria.

Wall, M. 2008. Quality of postharvest horticultural crops after irradiation treatment. *Stewart Postharvest Review.* **4** (2): 1.

White, I.M., and Elson-Harris, M. M. 1992. *Dacus* (Didacus) *ciliatus* Loew. **In**: Fruit Flies of Economic Significance: Their identification and Bionomics, pp. 329-331. CAB International, Wallingford, U.K.

World Trade Organisation. 1999. *Agreement on the Application of Sanitary and Phytosanitary Measures.* (http://www.wto.org/wto/legal/finalact.htm)

Yokoyama, V.Y. 1994, Fumigation. **In**: sharp, J. L and hallman, G. J (eds) quartantine treatments for pests of food plants. Westview press, boulder, Colourado, pp. 67-87.

Chapter 9

Pesticide Residues

Ahammed Shabeer T.P.

Scientist

Kaushik Banerjee

National Fellow

ICAR-National Research Centre for Grapes,
P.O. No.3, Manjri Farm, Pune – 412 307, Maharashtra, India

E-mail: shabsnrcg@gmail.com; E-mail: kbgrape@yahoo.com

1.0 Introduction

India is a major exporter of fresh fruits and vegetables. It ranks second in fruit and vegetable production in the world, after China. Amongst the fruits, the country ranks first in production of banana, papaya, mango, lemon and lime. India is also the largest producer of ginger and okra and ranks second in production of potato, onion, cauliflower, brinjal, cabbage, etc. The vast production base offers India tremendous opportunities for export. During 2011-12, India exported fruits and vegetables worth Rs. 4903 crores which comprised fruits worth Rs. 2156 crores and vegetables worth Rs. 2747 crores. Almost 25 percent of this export was to the European Union (EU) countries. Among the horticultural commodities, the major exported fruits are table grape, mango and pomegranate and vegetables like okra, green chilli, bitter gourd. The export of table grapes alone is about 70 million US dollars. Despite all these facts, India's share in the global food market is still <2 per cent only, which is mainly because of the challenging task in complying with the strict quality control standards in the importing countries like European Union, USA, etc. Compliance with international food standards is a prerequisite to gain a higher share of world trade. At the same time, there is a growing consciousness among Indian consumers for safe food due to recent Government initiatives through the Food Safety Standards Authority of India (FSSAI) on setting up of quality and safety standards for different food products at the national level.

Since many decades, a wide range of chemical pesticides are applied worldwide in agriculture. India being a tropical country, infestations of pests and diseases are frequent and regular use of plant protection chemicals is therefore inevitable to safeguard the crops. However, injudicious and indiscriminate use of pesticides is likely to have negative impact on human health through exposure to residues in food and drinking water besides via occupational, bystander, and residential exposure, etc., as well as through environmental contamination via wind drift, leaching, runoff, etc.[1] The general public is exposed to pesticides mainly through residues in agricultural food crops. Management of residues in fruits and vegetables is a difficult task and processing could be a way to minimize pesticide residue loads initially present in fresh fruits and vegetables. This chapter presents a general overview of pesticide residues, their analysis, dissipation, monitoring and management in fruits and vegetables, with special emphasis to India.

2.0 Pesticides and their Residues

2.1 Pesticides

The term **pesticide** has been defined by the World Health Organization (WHO) as any substance or mixture of substances intended for preventing or controlling any unwanted species of plants and animals and includes any substances or mixture of substances intended for use as a plant-growth regulator, defoliant or desiccant. It also includes any substance used for the control of pests during production, storage, transport, marketing or processing of food for man or animals or which may be administered to animals for the control of insects or arachnids in or on their bodies. It however, does not apply to antibiotics or other chemicals administered to animals for other purposes, such as to stimulate their growth or to modify their reproductive behaviour; nor does it apply to fertilizers. IUPAC also adopted the same definition for pesticide.

2.2 Pesticide Residue a Trade Barrier

Pesticide residue is any substance or mixture of substances in food for man or animals resulting from the use of a pesticide and includes any specified derivatives, such as degradation and conversion products, metabolites, reaction products and impurities which are considered to be of toxicological significance. The term also includes residues from unknown sources (*i.e.*, background residues) as well as those from known uses of any chemical in question.

Article 20 of the General Agreement on Tariffs and Trade (GATT) allows governments to act on trade in order to protect human, animal or plant life or health, provided they do not discriminate or use this as disguised protectionism. Sanitary and Phytosanitary Measures Agreement (SPS) allows countries to set their own standards.

However, regulations must be based on science, but "precautionary principle" may be applied. The Technical Barriers to Trade Agreement (TBT) tries to ensure that regulations, standards, testing and certification procedures do not create unnecessary obstacles. It encourages countries to recognize each other's procedures for assessing whether a product conforms to desired standards. Although TBT encourages removal of unnecessary obstacles in trade flow by avoiding over-stringent regulations, it simultaneously recognizes individual countries' rights to adopt their own standards which they might consider appropriate with respect to the life and health of human being, animal population, plants and environment.

Residue limits (MRLs) are the trade standards that are set in a way that there are no concerns for public health, especially with regard to vulnerable subpopula groups (as children and the unborn). Good Agricultural Practices (GAP) take into account the application of minimum quantities of pesticides necessary to achieve adequate pest control in such a manner that the amount of residues remaining in food are smallest possible. At the international level, the Codex Alimentarius Commission of the FAO/WHO decides the MRL. The Codex Committee on pesticide residues (CCPR) was formed by United Nations with primary mandate to establish MRL for pesticides in food. In India, the MRL of different pesticide-commodity combinations are prescribed under the Food Safety and Standard Authority of India Act, 2006 (FSSAI Act, 2006).

Although the Codex MRLs are applicable to all the signatory nations, individual countries are free to have their own MRL regulations. For agro-export, it is thus essential that the commodities should comply with the latest MRL regulations of the importing country. For example, the MRLs set by the European Union (EU) have to be complied for export of any food commodity (*e.g.* fruits and vegetables) from India to any of the EU member countries. Earlier, before September 2008, individual EU countries used to have their own MRLs. Since 1st September 2008, the MRLs have been harmonized across the member countries of the European Union, which has expanded the market access across the Europe and minimized trade barriers out of conflicting MRLs within the same territory. In addition to such Official MRLs, the Indian exporters aften have to comply with specific private standards imposed by supermarkets, where MRLs are typically at 50-70 per cent of the Official value rendering product compliance further challenging. Such private standards often raise apprehensions of combined toxicity in case of multiple pesticide detections in the same sample and even when the individual residue levels are well below the MRL.

During 2003, India faced major issues with respect to detection of pesticide residues in table grapes exported to the EU countries with several rapid alerts notified. Similar to grape, rapid alerts have also been received from the EU countries regarding the pesticide residues in other fruits (*e.g.* pomegranate) and vegetables (okra, chilli, etc.). If we assess the RASFF (Rapid Alert System for Food and Feed) notifications issued by the EU, for the third year in a row, there has been a continuous increase in RASFF notifications about pesticide residues. Out of the total 446 notifications for pesticide residues in 2012, only 34 were reported for food or feed originating from within the EU (RASFF Annual report, 2012) countries. Among this, alert issued to

Indian fruits and vegetables are alarming. No less than 77 notifications concerned okra from India which was included on the list of reinforced controls in 2011. Various residues were found on this product, often several compounds in single sample. Most frequently occurring pesticides included monocrotophos (31), acephate (22) and triazophos (15). Another frequently notified commodity from India is curry leaves which were added to the list of reinforced controls in 2010 with 21 notifications out of the 41 reported in 2012 due to triazophos at above MRL.

Trade issues related to detection of pesticides appear more complicated when unauthorized pesticides, *i.e.* the compounds which do not have official permission or label claim are detected. The number of authorized pesticides having a label claim are sometimes very few and therefore, the farmers do not have choice except use of old generic molecules with relatively higher mammalian toxicity and environmental stability. Current R&D efforts on pesticide residues are therefore mostly focused to replace such chemicals with those having a lower mammalian toxicity and safer environmental profile.

3. Pesticide Residue Analysis

Pesticide residue analysis involves several steps such as sampling, extraction, clean-up, analysis and data processing. The major steps are described below.

3.1 Sampling

Sampling is one of the crucial steps in pesticide residue analysis. It should be done in such a way that it is representative of the whole population and is without any bias. The sampling size and techniques, *i.e.* the size of sample, transport, storage, etc. might vary with commodities. The collected sample should be as homogeneous as possible before further processing. Random method of sampling most widely used and a detailed sampling procedure is given in the Codex guideline **CAC/GL 33-1999 entitled** "Recommended methods of sampling for the determination of pesticide residues for compliance with MRLs". An example of sample size required for plant products as per codex guidelines is given in Table 1.

Table 1: Plant Products: Description of Primary Samples and their Minimum Size

Commodity	*Example*	*Nature of Primary Sample*	*Minimum Size of Laboratory Sample*
Small sized fresh fruits and vegetables	Berries, Peas	Whole units or packages or units taken with sampling device	1 kg
Medium sized fresh fruits and vegetables	Apple, Brinjal Mango, Oranges	Whole units	1 kg (At least 10 units)
Large sized fresh fruits or vegetables	Cabbage, Cucumber Grape Bunches	Whole units	2 kg (At least 5 units)
Herbs	Fresh Parsley, Leafy vegetables	Whole units	0.5 kg

The samples should be transported to the laboratory in clean containers with robust packaging. Ventilated polythene bags are acceptable for most samples but low permeability nylon film bags must be used for samples to be analysed for residues of fumigants. Samples of commodities prepacked for retail sale should not be removed from their packaging before transport. Very fragile or perishable products (*e.g.* ripe raspberries) may have to be frozen to avoid spoilage and then transported in "dry ice" etc., to avoid thawing in transit. Samples that are frozen at the time of collection must be transported without thawing. Samples that may be damaged by chilling (*e.g.* bananas) must be protected from both high and low temperatures. Sample preparation, sample processing and sub-sampling to obtain analytical portions should take place before visible deterioration occurs. This is particularly important when the analytical result is considered to assess fitness for consumer intake. Canned, dried or similarly processed samples should be analysed within the stated shelf life. Where there is evidence that comminution (cutting and homogenisation) at ambient temperature has a significant influence on the degradation of certain pesticide residues, it is recommended that the samples are homogenised at low temperature (*e.g.* frozen and/or in the presence of "dry ice"). Where comminution is known to degrade residues (*e.g.* dithiocarbamates or fumigants) and practical alternative procedures are not available, the test portion should consist of whole units of the commodity, or segments removed from large units. For all other analyses, the whole laboratory sample (in most cases 1-2 kg) is thoroughly comminuted. As a thumb rule, residue analysis is carried out within the shortest time practicable after sampling to minimise any spoilage induced degradation of residues. In case, analysis is not immediately possible, the samples should be appropriately stored (usually at 0 °C) and analyzed within the shortest possible time.

3.2 Sample Extraction

Analysis of pesticides in food matrices (fruits and vegetables) is a difficult task, because of the complexity of the matrix and the low concentrations at which pesticide residues are usually present. Thus, despite advances in the development of highly efficient analytical instrumentations for their final determination, sample pretreatment remains an important part of obtaining accurate quantitative results. Many choices have been proposed for pretreatment and/or extraction of pesticide residues in foods. The extraction procedure usually involves sample homogenization with an organic solvent, alone or mixed with water or pH-adjusted water, using a homogenizer, blender, or sonicator. Some of the residue analysis techniques include solvent extraction (QuEChERS, Ethyl acetate method), supercritical fluid extraction (SFE), pressurized-liquid extraction (PLE), microwave-assisted extraction (MAE), matrix solid-phase dispersion (MSPD), solid-phase extraction (SPE), solid-phase micro extraction (SPME), and stir-bar-sorptive extraction (SBSE). Most of these techniques have resulted in new possibilities in sample treatment and advantages such as a substantial reduction in extraction time and incorporation into on-line flow-analysis systems. Each technique has its advantages and disadvantages and the choice should depend on the analytical problem.

Solvent extraction (SE) which may be followed by solid phase extraction (SPE) cleanup is still one of the most widely used techniques, mainly because of its ease of use, possibilities of analyte enrichment and wide ranging applicability. Such methods typically require a sample size of 50-100 g, which is extracted with solvents in the ratio of 1: 2, followed by cleanup by liquid-liquid partitioning and/or column chromatography. These methods have several disadvantages as they are laborious, time-consuming, expensive, and subject to problems arising from evaporation of large volumes of solvents and the disposal of toxic or inflammable solvents. To overcome the drawbacks mentioned above, new trends in sample treatment and miniaturization of time-tested SE have resulted in improved SE techniques that use much smaller amounts of organic solvent, enable more efficient extraction and faster cleanup ensuring easier laboratory automation and higher sample throughput. Good example for such SE methods are QuEChERS and ethyl acetate methods which because of increasing interest by many laboratories in recent years, are described in detail in the next section.

3.3 Sample Analysis

3.3.1 QuEChERS Method

QuEChERS means "quick, easy, cheap, effective, rugged and safe" method for the multiclass, multiresidue analysis of pesticides in fruits and vegetables. The QuEChERS approach takes advantage of the wide analytical scope and high degree of selectivity and sensitivity provided by gas and liquid chromatography (GC and LC) coupled with mass spectrometry (MS) for detection. A limited number of GC-amenable pesticides were evaluated in the original QuEChERS study. Although this version has been demonstrated to yield excellent results for hundreds of pesticides across a wide variety of commodities, subsequent experiments showed stability and recovery problems for some pesticides depending on pH of the matrix necessitating modifications of the original QuEChERS protocol by use of relatively strong acetate buffering conditions or weaker citrate buffering conditions in terms of ionic strength.

QuEChERS method went through extensive inter-laboratory trials entailing ≈50,000–100,000 data points for dozens of pesticides (both fortified and incurred samples) in different matrices and using different types of GC–MS and LC–MS conditions and instruments. These efforts successfully met statistical criteria for acceptability from independent scientific standards organizations, with the acetate-buffering version becoming AOAC official method 2007.01 and the citrate-buffering version being named European Committee for Standardization (CEN) Standard Method EN 15662.

QuEChERS is a quick and convenient replacement for LLE which furnishes high-quality results in a minimum number of steps and with low consumption of solvent and glassware. The original version of QuEChERS consists of extraction of homogenized samples (10 g) by hand-shaking or vortex mixing with 10 mL of acetonitrile (MeCN) to furnish a final extract sufficiently concentrated to avoid the need for solvent evaporation. Gram quantities of salts (4 g anhydrous magnesium sulfate, MgSO4, and 2 g sodium chloride, NaCl, a combination which affords well-defined phase separation without dilution with hazardous non-polar organic solvents) are then added to the sample, with mixing, to drive partitioning of the analytes between the aqueous and the solvent phases. After simple vortex mixing and centrifugation, which results in perfect physical separation of the phases, clean-up and removal of residual water is performed simultaneously by use of a rapid procedure, called dispersive solid-phase extraction (DSPE), in which a primary secondary amine (PSA) adsorbent and anhydrous MgSO4 are mixed with the sample extract. In dispersive SPE, the adsorbents are added directly to the extract without conditioning and the clean-up is easily performed by shaking and centrifugation. The latest procedure requires less time than traditional SPE and simultaneously removes residual water and many polar matrix components, for example organic acids, some polar pigments, and sugars.

The QuEChERS method has the advantages of high recovery, accurate results, high sample throughput, low solvent (no chlorinated solvents) and glassware usage, less labour and bench space, lower reagent costs, and high degree of ruggedness. Organic acids and other potential contaminants are removed during clean-up. The main disadvantage of QuEChERS is that for 1 g sample per mL of final extract the concentration of the extract is lower than the extracts obtained through traditional procedures, which uses much higher quantity (5-10 times or even more) of sample for extraction. Thus, the final extract must be concentrated to a greater extent to furnish the necessary sensitivity and to achieve the desired limits of quantification (LOQ). The success of this method also relies on application of highly sensitive and selective GC-MS and LC-MS techniques especially tandem mass spectrometry (MS/MS).

3.3.2 Ethyl Acetate Method

In a similar work flow as QuEChERS, ethyl acetate also provides high recoveries for most of the pesticides in fruits and vegetables. Ethyl acetate extract is more suitable for injection into GC-MS and for LC-MS analysis solvent exchange is carried out to methanol-water. It is especially suitable for the extraction of high-sugar commodities like grape, mango, etc. since sugar has limited solubility in ethyl acetate. Furthermore, substitution of acetonitrile by ethyl acetate can substantially reduce the input cost of analysis. Considering these advantages a multi-residue method in grape matrix was developed and validated. The sample preparation method follows extraction of homogenized grape samples (10 g) with 10 mL ethyl acetate (+10 g anhydrous sodium sulphate), cleanup by dispersive SPE with PSA (25 mg) and final analysis by LC–MS/MS after solvent-exchange to methanol–water (1: 1). Selection of ethyl acetate offered precise advantages over acetonitrile in minimizing the matrix components in the final extract and reducing the cost of analysis. This method provided an overall savings to the tune of around 30 per cent per sample when compared to the QuEChERS technique. Later this method was extended to other matrices and now this method is used by food testing laboratories in India for residue analysis fruits and vegetables for export.

3.3 Estimation

Once the sample preparation and extraction procedure is over, the sample has to be analysed by appropriate instrumentation techniques for the detection and quantification of the target analyte/analytes. Nowadays, by using GC combined with MS, simultaneous determination and confirmation of pesticide residues can be obtained with one instrument in one analytical run. In most cases, the sensitivity obtained with GC–MS is similar to that of classical GC detectors with the support of large volume injection and selective techniques like selected ion monitoring (SIM) and MS/MS. The commercial availability of atmospheric pressure ionization has caused a spectacular change in LC-MS applicability. Compared to traditional detectors, electrospray (ESI) and atmospheric pressure chemical ionization (APCI) in combination with MS instruments have increased the sensitivity of LC detection by several orders of magnitude. Moreover, with such progress in instrumentation techniques, the requirement of HPLC column switching techniques and extensive sample cleanup procedures unnecessary especially if tandem mass spectrometers are used and operated in the selected reaction mode (SRM). Due to the high level of selectivity by LC–MS/MS in the SRM, the signal-to-noise ratio increases distinctly and the full sensitivity range of LC–MS instruments can be utilized. Several researchers have reviewed the application of GC-MS/MS and LC-MS/MS technology for analysis of pesticide residues in fruits and vegetables.

4.0 Dissipation of Pesticides

After pesticides are applied to the crops, they may interact with the plant surfaces, get exposed to the environmental factors such as wind and sun and may be washed off during rainfall. The pesticide may be absorbed by the plant surface (waxy cuticle and root surfaces) and enter the plant transport system (systemic) or stay on the surface of the plant (contact). On the surface of a crop, a pesticide can undergo volatilization, photolysis, chemical and microbial degradation. Dissipation can be defined as the loss of pesticide residues in one compartment (*e.g.* plant tissue matrix) due to degradation and mass transfer to other compartments. Volatilization, photolysis and microbial degradation are the main processes by which pesticides get dissipated on fruits and vegetables. These processes are illustrated in Figure 1. All these processes can reduce the original pesticide's concentration load, but can also introduce some metabolites in the crops.

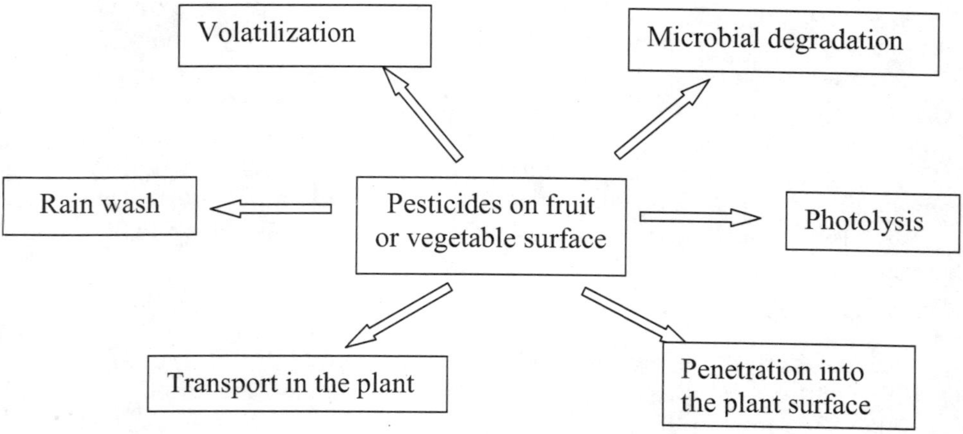

Figure 1: Dissipation of Pesticides on Fruit or Vegetables Surfaces

4.1 Dissipation Kinetics

A large number of kinetic models are used to describe the change in pesticide concentrations with time. This list of core models comprises the sample first-order model, and also a number of models that are able to describe bi-phasic degradation kinetics and lag phase models.

In case of pesticide dissipation or degradation in environmental systems, the above mentioned models represent simple and sensible approaches to mathematically describe the experimental data, and do not represent actual chemical reactions. Also, it may be noted that the more complex the dissipation/degradation pathway and the type of kinetics used, the more parameters the model will require, and the more data points are needed for adequate parameter estimation. Sample first-order kinetics (SFO) is expressed through exponential equation with only two parameters (total amount of chemical present at time t=0 and rate constant k). It assumes that the number of pesticide molecules is small relative to the number of factors micro-organisms and their enzymes or number of water molecules in case of hydrolysis. As a result, the rate of the change in pesticide concentration (dM/dt) is at any time directly proportional to the actual concentration remaining in the system. For SFO kinetics, the time for a decrease in the concentration by a certain percentage is constant throughout the experiment and independent of the initial concentration of the pesticide. First-order kinetics has also frequently been used to describe degradation in pesticide fate models. A number of publications have shown that simple first-order kinetics cannot adequately explain the degradation behaviour of pesticides in natural systems, where cannot the degradation pattern may follow a non-linear path. A non-linear two compartment first + first order model can adequately fit to the degradation pattern of many pesticides, and could also predict the half-life (DT_{50}) and preharvest interval (PHI) in a more realistic manner than a linear first-order model.

First-order kinetics and first + first order kinetic models can be represented by the following mathematical expressions:

First-order model: $[A]t = [A]1 \exp(-k1t)$

First + first-order model: $[A]t = [A]1 \exp(-k1t) + [A]2 \exp(-k2t)$

where, $[A]t$ is the concentration (mg $kg^{-}1$) of A at time t (days), $[A]1$ and $[A]2$ are the initial concentrations of A at time 0 degraded through first-order processes 1 and 2 and $k1$ and $k2$ are the degradation rate constants for 1 and 2. The units of k depend on the model used.

4.2 Dissipation Studies

Information on dissipation kinetics of pesticides in fruits and vegetables is a key aspect in residue management risk assessment practices. This is because human exposure to pesticides is predominantly caused by residues in agricultural crops grown for human and animal consumption. Dissipation kinetic models are used to determine the DT_{50} and PHI of pesticides in fruits and vegetables. The DT_{50}, which is the time at which the concentration of initial deposits reaches the 50 per cent level in the applied fruits or vegetables is calculated first order kinetics by the equation ln(2)

$\times$ k^{-1}. DT_{50} is an important parameter that signifies the speed of degradation. PHI is the time period (in days) required for dissipation of the initial residue deposits to below the MRL. DT_{50} and PHI data are very important in managing pesticide residues in any food commodity.

In India, several researchers have studied the fate of pesticides in/on fruits and vegetables and established crop-specific DT_{50} and PHIs. Several researchers investigated the dissipation of pesticides in a variety of fruits and vegetables in India and such data are successfully utilised for residue management in respective crops. Some recently reported dissipation studies on a variety of fruits and vegetables are summarised in Table 2.

Human exposure to pesticides is predominantly caused by residues in agricultural crops grown for human and animal consumption. Available information on pesticide dissipation in plants from experimental studies only covers a small fraction of possible combinations of substances authorized for use in agriculture. Additionally, aspects and processes influencing dissipation kinetics are still not fully understood. Therefore, Fantke and Juraske (2013) systematically reviewed 811 scientific literature sources providing 4513 dissipation half-lives of 346 pesticides measured in 183 plant species. They focused on the variability across substances, plant species and harvested plant components and finally discussed different substance, plant and environmental aspects influencing pesticide dissipation. Measured half-lives in harvested plant materials ranged from around 1 hour for pyrethrins in leaves of tomato and pepper fruit to 918 days for pyriproxyfen in pepper fruits under cold storage conditions. Ninety-five percent of all half-lives fall within the range between 0.6 and 29 days. These results emphasize that future experiments are required to analyze pesticide–plant species combinations that have so far not been covered and are highly relevant for assessment of human exposure.

5.0 Monitoring Pesticide Residues

The purpose of pesticide monitoring programs is to bring out the food safety status and ensure that the residue loads in fruits and vegetables do not exceed MRLs prescribed by the regulatory bodies. It also deters misuse of pesticides that could result in unexpected residues in food and facilitates implementation of good agricultural practices (GAP). The outcome of the residue monitoring programme is used by regulatory bodies for future and risk assessment exercises for public health. In most countries, the monitoring programs are organised by a single agency designated as the competent authority. The agency designs a monitoring plan based on the previous data available from dietary consumption and risk assessment exercises or pesticide usage in the available fruits and vegetables. Unfortunately, in India an ideal institutional frame work for this purpose is still lacking. Pesticide legislation in developing countries is generally lacking or not implemented and this also affects pesticide monitoring since it relies on legislation to be effective. In the European Union (EU) countries there is a coordinated programme under European Food Safety Authority (EFSA) for all the member countries as per the European Commission guidelines and the individual member states have their national residue monitoring programs.

Table 2: Dissipation of Half Life of Pesticides in Fruits and Vegetables

Commodity	Pesticide	Half Life (Days)	References
		a) Fruit Crops	
Apple	Kresoxim-methyl	4.58–4.77	Malhat *et al.*, 2013
	Trifloxystrobin tebuconazole	19.38–25.99	Oatyal *et al.*, 2013
	Propargite	1.66–2.61	Kumar *et al.*, 2005
	Emamectin benzoate	2.75–3.09	Wang *et al.*, 2012
Grape	Tetraconazole	4.5–9.5	Banerjee *et al.*, 2008
	Difenoconazole	4.5–5.5	
	Carbendazim	7.3–8.1	Kaushik Banerjee, 2005
	Mancozeb	4.6–5.7	
	Alpha-cyhalothrin	4.7–7.0	Banerjee *et al.*, 2006
	Methomyl	8.8–9.0	
	Azoxystrobin	5.4–11.2	Gajbhiye *et al.*, 2011
	Forchlorfenuron	5	Ugare *et al.*, 2013
	6-Benzyl aminopurine	3–3.5	
	Gibberellic acid	3: 2-3	
	Ethephon	12–15	
Mango	Imidachloprid	38	Bhattacherjee A. K., 2013
	Carbosulfan	7 d	
	Imidachloprid	1.86	Reddy *et al.*, 2013
	Heaxaconazole	1.97	
	Spirotetramat	3.3	Mohapatra *et al.*, 2012
	Imidacloprid	5.2	
		a) Vegetable Crops	
Brinjal	Thiacloprid	0.47 and 0.50	Sahoo *et al.*, 2013
	Profenofos	2.15–2.3	Mukherjee *et al.*, 2012
	Cypermethrin	10.91–1.86	
	Chlorpyrifos	3.27–3.10	
	Cypermethrin	2.19–3.27	
	Quinalphos	2-3	Pathan *et al.*, 2012
	Flubendiamide	0.62 and 0.54	Takkar *et al.*, 2012
Cabbage	Emamectin benzoate	1.34–1.72	Wang *et al.*, 2012
	Fipronil	3.21–3.43	Bhardwaj *et al.*, 2012
	Trichlorfon	1.80	Li *et al.*, 2011
	Quinalphos	4.8–5.3	Soudhamini and Deepa, 2013
	Fubendiamide	3.4–3.6	
	Flubendiamide	3.4 to 3.6	Paramashivam and Banerjee, 2013
	Thiacloprid	12.3–13.1	Dutta *et al.*, 2012

Contd...

Table 2–*Contd...*

Commodity	*Pesticide*	*Half Life (Days)*	*References*
	Spinosad	1.4 and 1.5	Singh and Battu, 2012
	metaflumizone	1.7–2.1	Chatterjee and Gupta, 2013
	Emamectin Benzoate	<5	Singh *et al.*, 2013
Cauliflower	Chlorantraniliprole	1.36	Kar *et al.*, 2013
	Cypermethrin	1.5–2.1	Gupta *et al.*, 2013
	Deltamethrin	2.9–3.3	
	Profenofos	2.6–3.0	
	Triazophos	2.2–2.6	
Chilli	Chlorpyriphos	4.43 and 2.01	Jyot *et al.*, 2012
	Cypermethrin	2.51 and 2.64	
	Trifloxystrobin	1.81 and 1.58	Sahoo *et al.*, 2012
	Tebuconazole	1.37 and 1.41	
	Deltamethrin	0.36–1.99	Pandher *et al.*, 2012
		0.38–2.06	
	Chlorpyriphos and	4.43 and 2.01	Jyot *et al.*, 2013
	Cypermethrin	2.51 and 2.64	
	Acetamiprid	2.24–4.84	Sanyal *et al.*, 2008
	Chlorfenapyr	2.93 to 2.96	Ditya *et al.*, 2010
Okra	Deltamethrin	2.09	Parmar *et al.*, 2012
	Alphamethrin	2.09	
	Deltamethrin in combination with triazophos	2.61 1.68	
	Ethion	1.27	
	Cypermethrin	2.59	
	Profenophos	1.88	
	Flubendiamide	4.7 to 5.1	Das *et al.*, 2012
	Spiromesifen	1.68 and 1.65	Raj *et al.*, 2012
	Imidacloprid	1.07 and 2.41	Banerjee *et al.*, 2012
	Beta-Cyfluthrin	1.98 and 3.30	
	chlorpyriphos	0.6	Samriti *et al.*, 2012
Tomato	Flubendiamide	0.72–1.32	Sharme and Parihar, 2013
	Thiacloprid	0.83–1.79	
	Flubendiamide	1.64–1.98	Paramashivam and Banerjee, 2013
	Bifenthrin	1.83–2.32	Chauhan *et al.*, 2012
	Chlorpyrifos	4.38–4.43	Rani *et al.*, 2013
	Mancozeb	3.76 and 4.14	Rani *et al.*, 2013
	Metalaxyl	1.29 and 0.41	

Several research publications or reports are available on pesticide residue monitoring in fruits and vegetables in India. In 2010, out of 34027 samples analyzed under the project "Monitoring of Pesticide Residues at National Level", 543 samples (1.6 per cent) were found above MRL. Commodity-wise break-up shows 3.1 per cent samples of vegetables, 0.8 per cent of fruits, were found to be above respective MRLs. In 2013, Kapoor *et al.*, screened around 250 samples, including fruits, fruit juices, and baby foods (50 samples each), vegetables (70 samples), and cereals (30 samples) collected from Lucknow, U.P., India, and analyzed for the presence of imidacloprid residues, out of which 22 per cent fruit samples showed the presence of imidacloprid, with 2 per cent detections being above MRL. In case of the vegetables, out of the 24 per cent positive samples, 5.71 per cent had residues above the MRL. In 2012, fruits/ vegetables were collected from the Nalgonda district of Andhra Pradesh for pesticide residues (monocrotophose, quinalphose and carbofuran). The data revealed that pesticide residues in processed as well as unprocessed samples were below detectable limits (BDL) in all the samples. Further investigation on the distribution of pesticide residues in commonly used vegetables form Hyderabad reported the mean concentration of chlorpyrifos in eggplant (24.02 µg/kg), cabbage (10.55 µg/kg), cauliflower (2.85 µg/kg), tomato (178.87 µg/kg) and okra (2.49 µg/kg) differed significantly ($p < 0.002$). Similarly, the mean concentration of triazophos in eggplant (0.863 µg/kg), cabbage (2.21 µg/kg), cauliflower (0.491 µg/kg), tomato (3.01 µg/kg) and okra (2.49 µg/kg) differed significantly ($p < 0.007$). A similar trend was observed for acephate, fenitrothion and phosalone. The presence of target pesticides was found in the range of 0.016-50.85 µg kg^{-1}. Similarly, distribution of pesticide residues in vegetable samples from Andaman Island revealed 34 per cent positive for pesticide residues. Among the OC compounds, α-endosulfan, β-endosulfan, and endosulfan sulphate; synthetic pyrethroids, such as á-cypermethrin, fenvalerate I, fluvalinate I, deltamethrin, and dikudd-cyhalothrin and organophosphates such as chlorpyrifos, profenofos, monocrotophos, triazophos, ethion, dimethoate, and acephate were detected. Out of the positive samples, 15.3 per cent were found to contain residues exceeding the prescribed MRLs. Consequently, it was recognized that regularly consumed marketable fruits and vegetables contain measurable quantities of OC, OP and pyrethroid pesticides. Fifty Vegetable samples (beans, brinjal, cabbage and carrot) grown in Kolar district of Karnataka, India were analysed for 20 pesticide residues by GC equipped with electron capture and flame thermionic detectors (ECD and FTD) and all the samples were found contaminated with frequency of OC detections (97 per cent) dominated followed by OPs (83 per cent) and pyrethroids (60 per cent). Around 58 per cent of the samples were found to contain the residues of these insecticides above their respective MRLs.

Out of the 75 tomato samples tested, 26 (35 per cent) were contaminated and 4 per cent had residues above MRLs. Among the 80 brinjal samples, 46 (56 per cent) contained residues and 4 per cent of samples had residues above MRLs. Bupander Kumar, 2012 studied the distribution of OC residues in fruits and vegetables from India. The total concentration of OC pesticides (OCPs) ranged from 0.29 to 106.65 µg.kg^{-1}. Vegetables had residue levels of OCPs much below the recommended MRLs set by the European Commission and Indian Government, indicating minimal risk to the consumers. Pesticide residues in market samples of vegetables from Lucknow district of Utter Pradesh was investigated and twenty-three pesticides were detected (a list of 48 pesticides were tested) in the samples in the range of 0.005–12.35 mg kg^{-1}. The detected pesticides were: Σ-HCH, Dicofol, Σ-Endosulfan, Fenpropathrin, Permethrin-II, β-cyfluthrin-II, Fenvalerate-I, Dichlorvos, Dimethoate, Diazinon, Malathion, Chlorofenvinfos, Anilophos, and Dimethachlor. In some vegetables like radish, cucumber, cauliflower, cabbage, and okra, the detected pesticides (Σ-HCH, Permethrin-II, Dichlorvos, and Chlorfenvinfos) were above the MRL (PFA 1954). However, in other vegetables the level of pesticide residues was either below detection limit or MRL.

In view of the detection of pesticide residues these findings several rapid alerts were being issued by the importing countries and the situation is precarious enough where import of certain fruits and vegetables from India could be banned. More than the presence of pesticide residues above MRLs, a serious fact revealed was the detection of non-recommended pesticides in vegetables. Moreover, the pesticides which have a label claim are mostly old generic molecules exhibiting a higher mammalian toxicity. These molecules do not have any label claim and beckon an immediate replacement/substitution with chemicals having a low mammalian toxicity and green chemistry. Thus, extensive data generation regarding the bioefficacy, residues dissipation, dose standardization, PHI and application technology of these molecules needs to be conducted at research institutions for developing a safe package of practice nucleating upon rigorous stringent and methodical research activities.

6.0 Risk Assessment

Risk assessment of chemicals is described as a process intended to calculate or estimate the risk to a given target organism, system or (sub) population, including identification of attendant uncertainties, following exposure to a particular agent taking into account the inherent characteristics of the agent of concern as well as the characteristics of the specific target system (OECD 2003). Risk assessment process includes four steps: hazard identification, hazard characterisation (dose-response assessment), exposure assessment and risk characterisation.

6.1 Hazard Identification

It is the first step in risk assessment and it involves the identification of the type and nature of adverse effects that an agent has as inherent capacity to cause in an organism, system or (sub) population (OECD 2003). It is generally performed before a pesticide is approved for usage in agriculture or other areas and the information on hazards posed is readily available from the pesticide registering bodies and on their websites for public usage. Most of such information is also available from international organisations such as JMPR, OECD and EC. For hazard identification, studies are carried out in test organisms (microbial, cells or animals) and the exposure level is increased until an adverse effect is produced. The highest dose of a pesticide that does not cause detectable toxic effects on the test organisms is called the no-observed-adverse-effect-level (NOAEL) and is expressed in terms of mg per kg of body weight per day (WHO 1997). It is used in calculation of the acceptable daily intake (ADI).

6.2 Hazard Characterisation

Hazard characterisation is the qualitative and, wherever possible, quantitative description of the inherent properties of an agent or situation having the potential to cause adverse effects. This should, wherever possible, include a dose response assessment and its attendant uncertainties (OECD 2003). Hazard characterisation involves comparing the pesticide exposure concentration with the ADI. The ADI is the estimate of the amount of a substance in food (mg/kg body weight/day) that can be ingested daily over a lifetime without appreciable health risk to the consumer (WHO 1997). ADI is calculated by dividing the NOAEL for animal studies with an uncertainty factor of 100 to convert to a safe level for humans.

6.3 Exposure Assessment

It is the evaluation of the concentration or the amount of a particular agent that reaches a target organism, system or (sub) population in a specific frequency for defined duration (OECD 2003). The potential intake or consumption of pesticide residues is divided by the body weight and compared to ADI.

Exposure = (Concentration of pesticide residue x Food consumed)/body weight

Dietary exposure assessment is very important since the food consumption pattern varies from country to country. So, as much as possible, the food consumption data used for dietary exposure should be the average daily consumption of the particular commodity in the targeted population. If such information is not available, the WHO data on average daily consumption of food can be used although it could be too much generalized.

6.4 Risk Characterisation

The qualitative and, wherever possible, quantitative determination, including attendant uncertainties, of the probability of occurrence of known and potential adverse effects of an agent in a given organism, system or (sub) population, under defined exposure conditions (OECD 2003). The international estimated daily intake (IEDI) has been used to characterize the risk of pesticides and it is expressed as:

$$IEDI = \Sigma\, STMR \times E \times P \times F$$

where,

STMR = supervised trial median residue level

E = Edible portion

P = processing factor

F = consumption of the food commodity

When the IEDI is more than the ADI, the food involved is considered a risk to the concerned consumers.

In India, an institutional frame work is being established under the aegis of FSSAI for the risk assessment of pesticide residues in food commodities.

7.0 Pesticide Residue Management

Considering the national and international food safety issues, it is very important to adopt residue management strategies to tackle the pesticide residue problems in fruits and vegetables and thereby increase the export of these commodities and also to ensure safety to domestic consumers. Some strategies for pesticide residues management in fruits and vegetables are described below

7.1 Processing

Fruits and vegetables like other foods often pass through culinary and food processing treatments before they are consumed. The effects of these culinary and food processing techniques have been investigated by various researchers and they have been found to reduce the pesticide residue levels except in cases where there is concentration of the product like in juicing, drying and oil production. Some toxic metabolites may be produced during processing treatments, especially thermal processing. However, the consumers can still be encouraged to employ those processing methods that reduce pesticide residues. Food processing studies often results in transfer factors or food processing factors (PF) of the pesticide residue in the transition from raw agriculture commodity to the processed product. These processing factors are expressed as the concentration of pesticide after processing divided by the concentration before processing. Some processing factors are available in public literature while others are only available from the pesticide registering bodies. Processing studies have become a part of pesticide registration requirements. Some processing techniques and average processing factor are summarized in Table 3. Kaushik *et al.* (2009) reviewed the effect of processing on dissipation of pesticide residues in fruits and vegetables.

Table 3: Processing Techniques and Average Processing Factor

Processing Technique	*Processing Factor*
Baking	1.38
Blanching	0.21
Boiling	0.82
Canning	0.71
Frying	0.1
Juicing	0.59
Peeling	0.41
Washing	0.68

Source: Adopted from: Keikotlhaile, B. M and Spanoghe P., 2011.

The effect of the above processing techniques on pesticide residue dissipation are given below.

7.1.1 Baking

Baking is the technique of prolonged cooking of food by dry heat normally in an oven. It is primarily used for the preparation of bread, cakes, pastries and pies, tarts, and quiches. It is also used for the preparation of baked potatoes; baked apples; baked beans. During baking when substrates undergo heating, the loss of pesticide residues may be through some physicochemical processes, *e.g.* evaporation, co-distillation and thermal degradation which may vary with the chemical nature of the individual pesticides. During the process the water contained in the tissue could entrain pesticide molecules (co-distillation) while heat causes evaporation and degradation.

7.1.2 Blanching

Hot water blanching increases pesticide removal and may hydrolyze substantial fractions of non-persistent compounds. The blanching operation (after peeling) of the contaminated potatoes (at level of 1 ppm) resulted in reduction of the residues by 28.3 per cent, 22.9 per cent, 26.0 per cent, 47.3 per cent, 46.3 per cent and 45.9 per cent for Hexachlorobenzene, lindane, pp′-DDT, dimethoate, pirimiphos-methyl and malathion, respectively. Blanching affects organophosphorous pesticides more than organochlorine, which withstands 100 °C. This may be due to the high stability of organochlorines to heat treatment. Tomatoes fortified with 25 ppm of parathion were processed by blanching which resulted in 50 per cent reduction in parathion residue level.

7.1.3 Boiling

Pesticide residues might be deteriorated and/or degraded during cooking. In particular, pesticide compounds with functionalities such as carbamate, amide, urea, thiocarbonyl, and imino group are readily hydrolyzed in the presence of trace amounts of acid and/or base during heating or boiling. Processing of spinach for 66 min at 252 °F reduced the pesticides as diazinon 58 per cent, azinphos-methyl 100 per cent, malathion 96 per cent, methyl parathion 100 per cent and carbophenothion 17 per cent. Some studies show an increase of ethylenethiourea (ETU) formation when cooking plant materials treated with ethylene bisdithiocarbamates (EDBCs) fungicides.

7.1.4 Canning

Canning in its various forms combines elements of washing, peeling, juicing, cooking and concentration. Processing whole tomatoes with vinclozolin residues of 0.73 mg/kg subsequently resulted in residue detections in canned juice, puree and ketchup to the extent of 0.18, 0.73 and 0.22 mg/kg, respectively. A canning process that did not include peeling for cherries removed 95 per cent of the tetrachlorvinphos residues (4.3 ppm).

7.1.5 Frying

Processes involving heat can increase volatilization, hydrolysis or other chemical degradation and thus reduce residue levels. Blanching and frying of eggplant for 5 min completely removed the profenofos residues which were initially present at level of 0.27 ppm. Frying (after peeling) affected organophosphorous residues more than organochlorine as the percent reduction of organophosphorous ranged between 49 per cent and 53 per cent.The level of reduction ranged between 30.1 and 35.3 per cent for the organochlorines.

7.1.6 Washing

Washing with water and various chemical solutions for domestic and commercial use are necessary to decrease the intake of pesticide residues. Washing is the most common form of processing which is a preliminary step in both household and commercial preparation. Loosely held residues of several pesticides are removed with reasonable efficiency by varied types of washing processes. Bitter gourds treated with endosulfan sprays received initial deposits of 18.97 ppm and 26.01 ppm which were removed to the extent of 59.05 per cent and 42.66 per cent respectively upon 30 s of washing. Similarly, 75 per cent of the residues of azoxystrobin initially present on grape (0.49–1.84 ppm) were removed by washing. Captan residues in apples when washed for 10-15 s with continuous hand rubbing were 50 per cent lower than in those apples that received no postharvest washing (25.5–5100 ng/g).

7.1.7 Peeling

Peeling of the fruits and vegetables can physically remove the surface residues present. Freezing as well as juicing and peeling are necessary to remove the pesticide residues in the skins. A decrease of about 77 per cent of the initial procymidone residue level (0.86 ppm) on tomatoes was observed after peeling.

While, food processing usually results in a decrease in pesticide or contaminant levels, in some cases, residue levels may increase in the final product due to concentration factors of raw commodities in the process of the final product. This concentration effect can be related with water removal for example in tomatoes used for tomato ketchup or in the production of drying fruits such as raisins and prunes. In such cases, a processing factor (PF) of greater than one is observed. A PF value of >1 in raisin was observed for cypermethrin and azoxystrobin.

7.2 Preharvest Interval (PHI)

PHI is most important in field level management of pesticide residues. On the basis of the MRL, the PHIs of pesticides are calculated. PHI is also termed as the safe waiting period, which is the minimum time in days that must be provided between last application of a pesticide and harvesting of the produce so that its residue level at harvest reaches below the MRL. Estimating PHI of a pesticide ideally involves multi-location field trails wherein the pesticides are applied following the guidelines of GAP. Although first order rate kinetics is largely used for estimating PHI, non-linear kinetics, *e.g.* first + first order often governs the dissipation of pesticides, where the residues dissipate with time following a non-linear relationship. In most of the dissipation studies described in the above section, the PHI was also established for respective pesticide in studied crops. When the PHIs are strictly followed, the pesticide residue loads are expected to dissipate to below MRL levels in the harvested produce.

7.3 Spraying Techniques

Appropriate spraying technology improves field coverage with lesser quantity/ volume and therefore minimizes the actual amount of pesticides. Most of the pesticides which are sprayed in the field don't reach the target. So, the use of sophisticated spraying technologies like electrostatic sprayers as a precision farming technology can reduce the amount of pesticide usage to some extent.

7.4 Integrated Pest Management

To some extent, IPM can reduce the usage of chemical pesticides in the field. As much as possible, use environmental friendly startegies and involve the less persistent synthetic pesticide in the IPM package.

7.5 Field Level Management

The major things to take care at field level management are:

i) Use only the recommended pesticides

ii) Do not spray banned as well as non-recommended chemicals

iii) Strictly adhere to the recommended dose of application including the recommended volume of water to be sprayed per hectare per acre.

iv) Strictly maintain recommended PreHarvest Interval to minimize consumer risk regarding detection of residues

v) Follow Integrated Pest Management practices involving minimal usage of highly toxic chemicals

vi) Sequencing the pesticides in relation to their persistence particularly in the last 2 months period before harvest

vii) Adoption of Bio-control measures during last month before harvest

viii) "Combination Products" containing two or more active ingredients in the above list can be used. However, the dose of the individual active ingredient should not exceed the recommended dose.

7.6 Regulatory Control

Knowingly or unknowingly farmers keep applying banned or restricted use chemicals for pest and disease management. It is extremely important to replace the usage of highly toxic chemicals with those having safer mammalian and environmental toxicity profiles. In many cases, the detection of the pesticide residues in fruits and vegetables include restricted use chemicals or those which are not having label claim. Strict regulatory control need to be established to minimize the use of such restricted or unauthorized (without official label claim) chemicals. This can be achieved through extension programmes involving all the stakeholders *e.g.* farmers, state government agricultural/horticultural department officials and exporters under the overall supervision of Food Safety and Standard Authority of India and Central Insecticide Board.

8.0 Conclusion

Consumer awareness is growing with respect to the right to have safe food. No one in the society is ready to tolerate pesticide residues in the food intended for his/her nurture. On the other hand, being in tropical belt we do not have the luxury of avoiding the use of agrochemicals in pest management. It is the dose and time of application, which differentiate the safe and unsafe use of pesticides. If the recommendations of the GAP and PHI are sincerely followed, it is definitely possible to minimize the residue load of agrochemicals in fruits and vegetables at field level, which in turn, will ensure safety to the health of the consumers and ecosystem as a whole. The monitoring data reveals that the market samples of fruits and vegetables have residues above the prescribed MRLs. Processing can be used as the best tool for postharvest management of pesticide residues. Risk assessments associated with pesticide residues need to be scientifically executed in a structured manner. A well-structured residue monitoring system with an institutional frame work can possibly ensure field-level traceability and management of agrochemical residues in fruits and vegetables through implementation of good agricultural practices.

9.0 Important References

Alder, L., Greulich, K., Kempe, G. and Vieth, B., 2006. Residue analysis of 500 high priority pesticides: better by gc–ms or lc–ms/ms? *Mass Spectrometry Reviews*, 25: 838– 865

Anastassiades, M., Lehotay, S.J., Stajnbaher, D. and Schenck, F.J., 2003. Fast and easy multi-residue method employing acetonitrile extraction/partitioning and "dispersive solid-phase extraction" for the determination of pesticide residues in produce. *J. AOAC Int.*, 86: 412-431.

Anastassiades, M., Scherbaum, E., Tas¸ B., delen, Stajnbaher, D., Ohkawa, H., Miyagawa, H. and Lee, P.W., 2007 (Eds.). *Crop Protection: Public Health, Environmental Safety*. Wiley-VCH, Weinheim, Germany, p. 439.

Banerjee, K., Oulkar, D.P., Dasgupta, S., Patil, S.B., Patil, S.H. and Savant, R., 2007. Validation and uncertainty analysis of a multi-residue method for pesticides in grapes using ethyl acetate extraction and liquid chromatography–tandem mass spectrometry. *Journal of Chromatography* A 1173: 98–109.

Farhath, S. and Shobha, S., 2012. Pesticide residues in selected fruits/vegetables of Nalgonda district of Andhra Pradesh. *J. Res. ANGRAU*, 40(3): 64-66.

Gowda, S.R.A. and Somashekar, R.K., 2012. Evaluation of pesticide residues in farm-gate samples of vegetables in Karnataka, India. *Bulletin of Environmental Contamination and Toxicology*, 89(3): 626-632.

Herna´ndez, F., Sancho, J.V. and Pozo, O.J., 2005. Critical review of the application of liquid chromatography/mass spectrometry to the determination of pesticide residues in biological samples. *Anal. Bioanal. Chem.*, 382: 934–946.

Jha, Rishabh, 2013. Multi-residue pesticide analysis of major pesticides in marketable fruits and vegetables. *Masters Thesis*, Indian Institute of Science Education and Research Kolkata, URL: http://www.iiserkol.ac.in

Kaushik, G., Satya, S. and Naik, S.N., 2009. Food processing a tool to pesticide residue dissipation: A review. *Food Research International*, 42: 26–40.

Keikotlhaile, B.M. and Spanoghe, P., 2011. *Pesticide Residues in Fruits and Vegetables; Pesticides - Formulations, Effects, Fate*, Edited by Prof. Margarita Stoytcheva, p. 243-252.

Kumar, B. and Mukherjee, D.P., 2012. Organochlorine residues in vegetables. *International Journal of Vegetable Science,* 18(2): 121-136.

Kumari, B.R., Ranga Rao, G.V.R., Sahrawat, K.L. and Rajasekhar, P., 2012. Occurrence of insecticide residues in selected crops and natural resources. *Bulletin of Environmental Contamination and Toxicology*, 89(1): 187-192.

Lehotay, S.J., Hiemstra, M., van Bodegraven, P. and de Kok, A., 2005. Validation of a fast and easy method for the determination of residues from 229 pesticides in fruits and vegetables using gas and liquid chromatography and mass spectrometric detection. *J. AOAC Int.*, 88: 595-614.

Lehotay, S.J., 2002. Application of gas chromatography in food analysis. *Trends in Analytical Chemistry,* 21(9,10): 686-697.

OECD, 2003. Descriptions of selected key generic terms used in chemical hazard/risk assessment. *OECD Series on Testing and Assessment*, No. 44.

Peter Fantke, P. and Juraske, R., 2013. Variability of pesticide dissipation half-lives in plants. *Environ. Sci. Technol.*, 47(8): 3548–3562. DOI: 10.1021/es303525x.

RASFF Annual Report, 2012. The Rapid Alert System for Food and Feed (ec.europa.eu/food/food/rapidalert/./rasff_annual_report_2012_en.pd).

Sinha, S.N., Rao, M.V.V and Vasudev, K., 2012. Distribution of pesticides in different commonly used vegetables from Hyderabad, India. *Food Research International,* 45(1): 161–169.

Sinha, S.N., Rao, M.V.V and Vasudev, K., 2012. Quantification of organophosphate insecticides and herbicides in vegetable samples using the "Quick Easy Cheap Effective Rugged and Safe" (QuEChERS) method and a high-performance liquid chromatography-electrospray ionisation-mass spectrometry (LC–MS/MS) technique. *Food Chemistry*, 132(3): 1574-1584.

Srivastava, A.K., Trivedi, P., Srivastava, M.K., Lohani, M.L. and Srivastava, L.P., 2011, Monitoring of pesticide residues in market basket samples of vegetable from Lucknow City, India: QuEChERS method. *Environmental Monitoring and Assessment*, 176(1-4): 465-472.

Swarnam, T.P. and Velmurugan, A., 2013. Pesticide residues in vegetable samples from the Andaman Islands, India. *Environmental Monitoring and Assessment*, 185(7): 6119-6127.

Upasana Kapoor, U., Srivastava, M.K., Srivastava, A.K., Patel, D.K., Garg, V. and Srivastava, L.P., 2013. Analysis of imidacloprid residues in fruits, vegetables, cereals, fruit juices, and baby foods, and daily intake estimation in and around Lucknow, India. *Environmental Toxicology and Chemistry*, 32(3): 723-727.

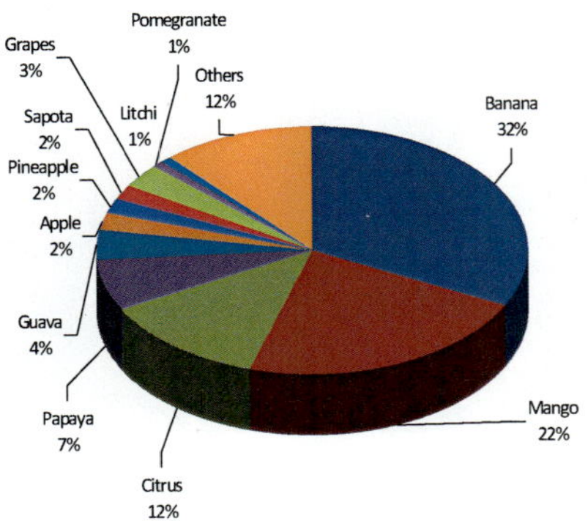

Production Share of Major Fruits in India (2012-13) (p. 3)

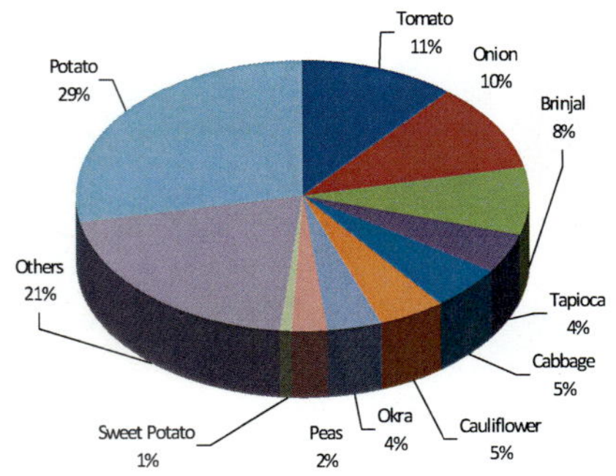

Production Share of Major Vegetables in India (2012-13) (p. 4)

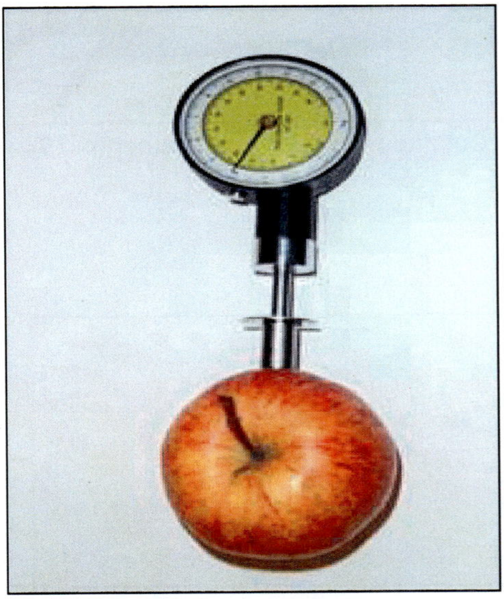

Fruit Firmness Tester (p. 50)

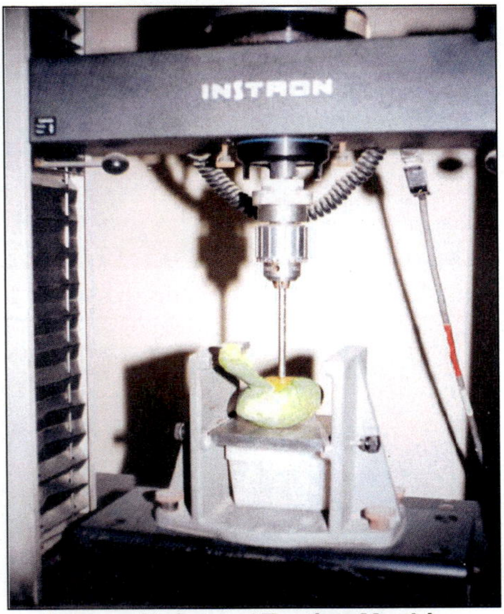

Instron Universal Testing Machine (p. 50)

30 Days after Flowering

40 Days after Flowering

50 Days after Flowering

Different Stages of Maturity in Capsicum (p. 53)

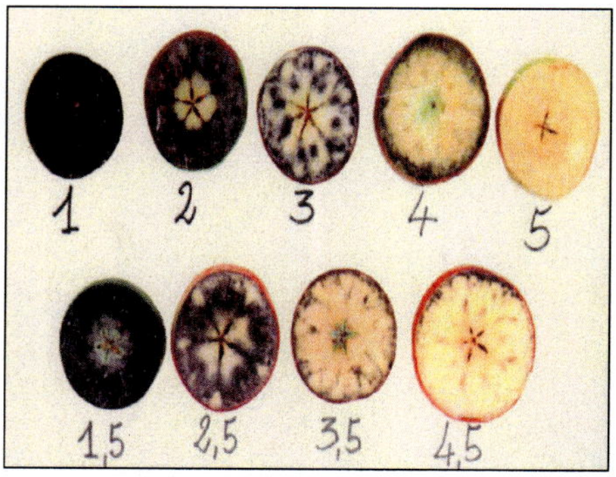

Starch-Iodine Score for Apple Maturity (p .54)

Lime Harvester * (p. 79)

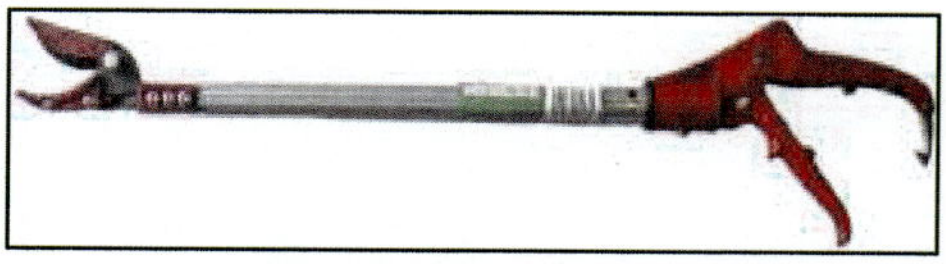

Cut and Hold Harvester/Pruner (p. 79)

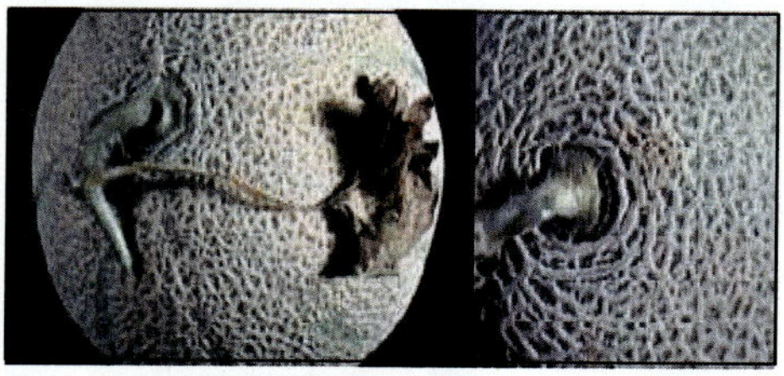

Maturity of Muskmelon (p. 64)

Green Stage **Red Stage** **Light Red Stage** **Pink Stage** **Turning Stage** **Breaker Stage**

Maturity Stages of Tomato (p. 65)

Mango Harvester (p.78)

Sapota Harvester (p.78)

Tripod Ladder (p. 80)

Self Propelled Hydraulically Operated Man Positioning (p. 80)

Tractor Trailor Mounted Platform (p. 80)

Light Pole Harvester for Oilpalm and Coconut (p. 81)

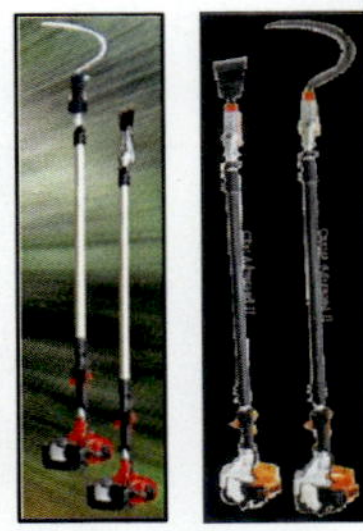

Engine Operated Harvesters for Oilpalm (p. 81)

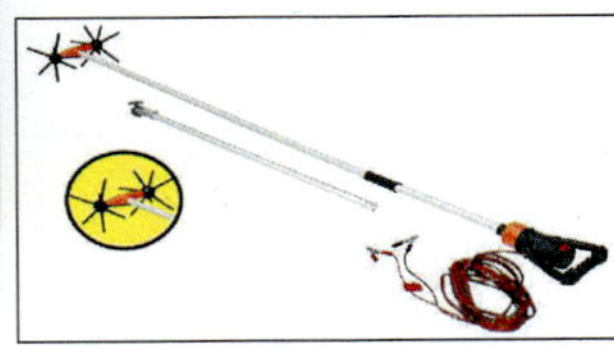

(a) Electrical shaker harvester (p. 81)

(b) Olive shaker harvester

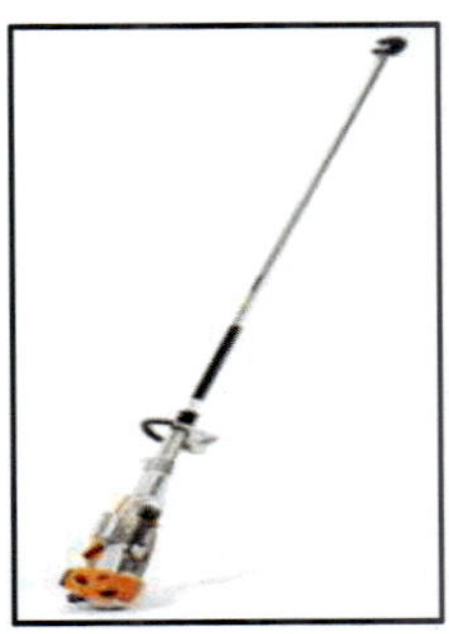

(c) Shaker harvester for branch

(d) Shaking olive branch

Shaker Harvesters (p. 81)

Shaker Harvester for Apple (p.82)

(p.83)

Self Propelled Shaker Harvester with Inverted Umbrella Collection Apron and Belt Conveyor (p. 83)

Self Propelled Tree Shaker (p. 83)

Spiked Drum Shaker Harvester (p. 84)

Canopy Shaker Harvester for Orange (p. 84)

Grape Harvester (p .86)

Harvesting Pineapples Manually and Putting on Conveyor (p .87)

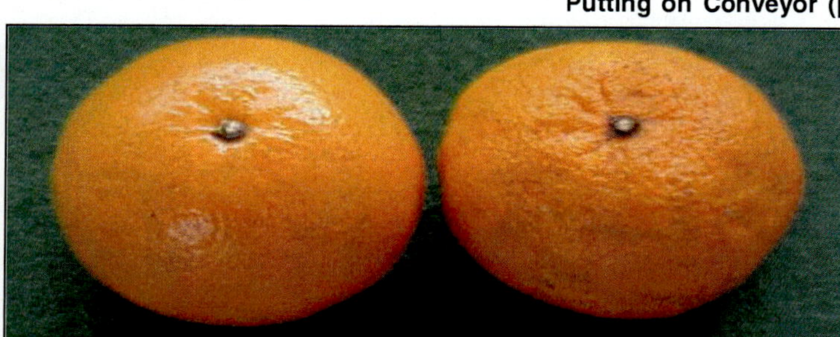

Wax Coating **Without Coating**

Effect of Wax Coating on Appearance of Kinnow Fruits (p.100)

CIPC Treated Potato — Untreated Potatoes

Effect of Postharvest Application of CIPC on Sprouting of Potato (p. 103)

Effect of Shade on Quality of Capsicum (p. 108)

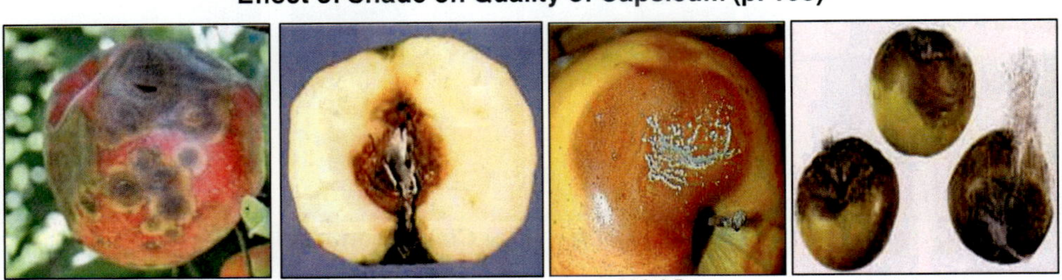

(a) Bitter Rot (b) Core Rot (c) Soft Rot (d) Whisker Rot

Postharvest Diseases of Apple (p. 114)

(a) Anthracnose (b) *Botryodiplodia* Rot (c) Cigar End Rot

Postharvest Diseases of Banana (p. 116)

(a) *Alternaria* Fruit Rot

(b) *Aspergillus* Rot

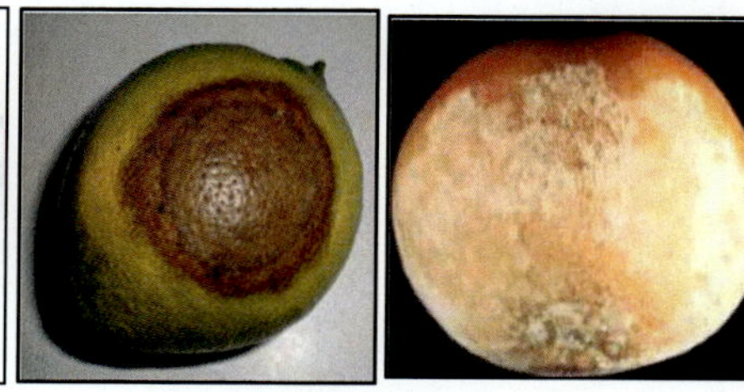

(C) BROWN ROT

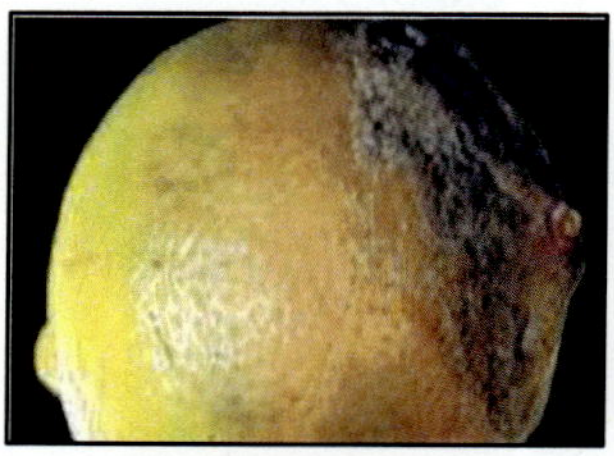

(d) *Colletotrichum* Stem End Rot

(e) *Penicillium* Fruit Rot

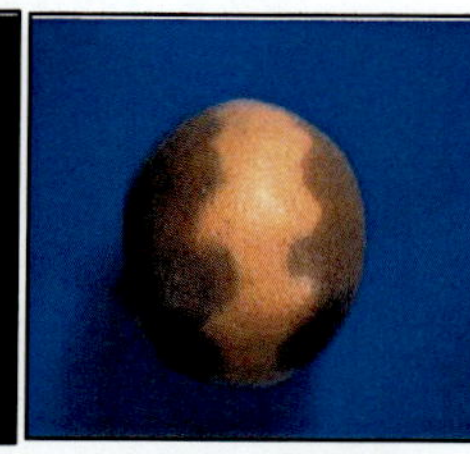

(f) Stem End Rot

Postharvest Diseases of Citrus (p. 119)

(a) Bitter Rot

(b) Blue Mould Rot

(c) *Cladospori*um Rot

(d) Grey Mould Rot

Postharvest Diseases of Grapes (p. 120)

(a) *Colletotrichum* Fruit Rot

(b) Fruit Canker

(c) *Phoma* Fruit Rot

Figure 5: Postharvest Diseases of Guava (p. 122)

(a) Anthracnose (b) Black Mould Rot (c) Stem End Rot

Postharvest Diseases of Mango (P. 123)

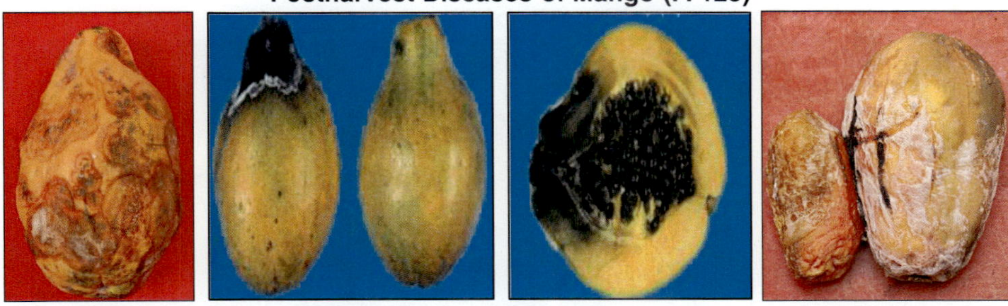

(a) Anthracnose (b) *Phomopsis* Fruit Rot (c) *Rhizopus* Rot

(http://agritech.tnau.ac.in)

Postharvest Diseases of Papaya (p.126)

Stem End Rot of Pineapple (p. 127)

(a) Black Mould (b) Dry Rot (c) Heart Rot

Postharvest Diseases of Pomegranate (p. 129)

(a) Anthracnose

(b) Grey Mould

Postharvest Diseases of Strawberry (p. 130)

(a) Anthracnose (http:// www.cals.ncsu.edu)

(b) Cottony Leak (https://extension. udel. edu)

(c) Grey Mould-Rot (http:// mtvernon.wsu.edu)

(d) Pod Rot (https:// extension.udel.edu)

Postharvest Diseases of Beans (p.132)

(a) Black Rot (http://vegetablemdonline. ppath.cornell.edu)

(b) Grey Mould (http://www.farminguk.com)

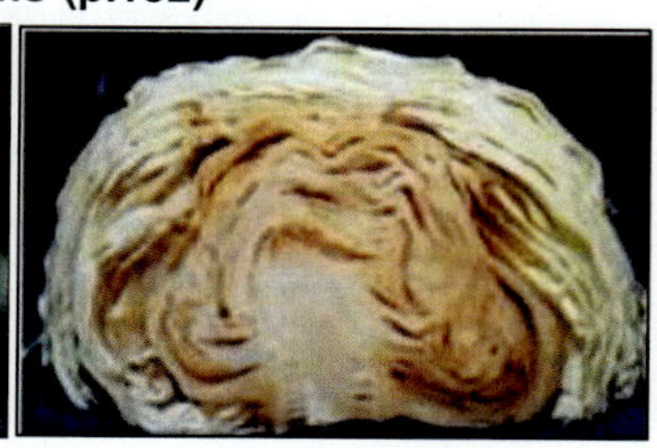

(c) Soft Rot (http://www.agroatlas.ru)

Postharvest Diseases of Cabbage (p. 133)

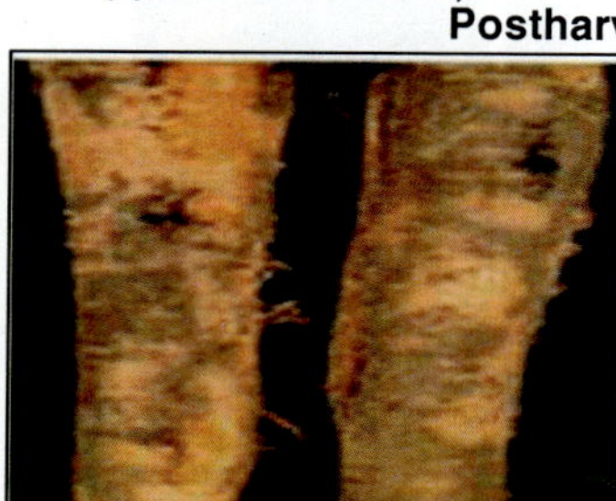

(a) *Alternaria* Rot

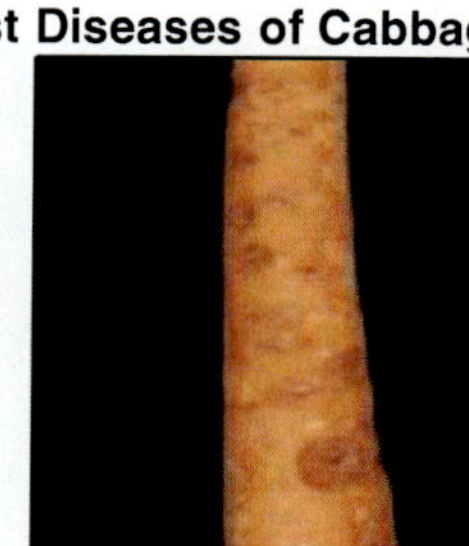

(b) Bacterial Soft Rot (https://

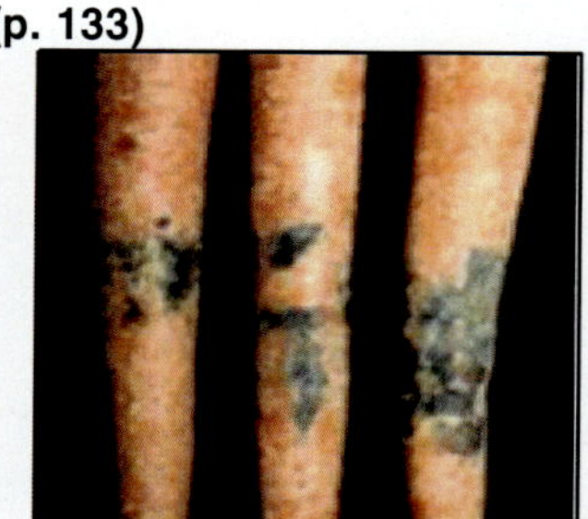

(c) Black Root Rot (https://

Postharvest Diseases of Carrot (p. 134)

(d) Cottony Soft Rot
(http://agritech.tnau.ac.in)

(e) Crown Rot
(http://agritech.tnau.ac.in)

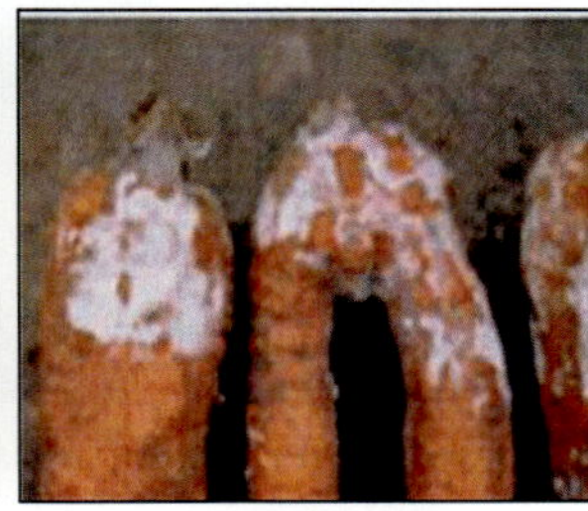

(f) Dry Rot
(http://agritech.tnau.ac.in)

Postharvest Diseases of Carrot (p. 134)

(a) Black Mould Rot
(http://mtvernon.wsu.edu/path_team)

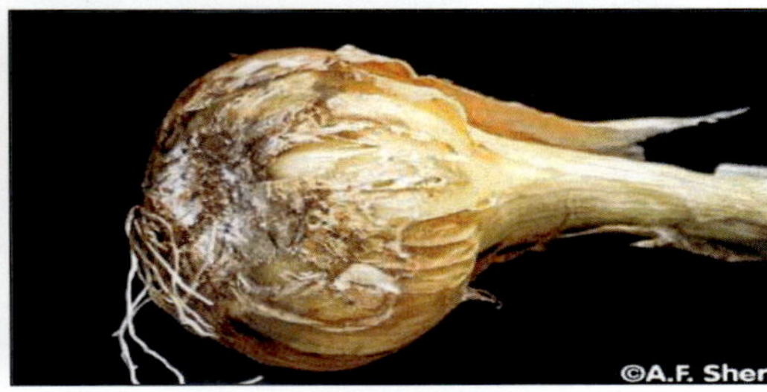

(b) Bulb Rot
(http://vegetablemdonline.ppath.cornell.edu/)

Postharvest Diseases of Onion (p.136)

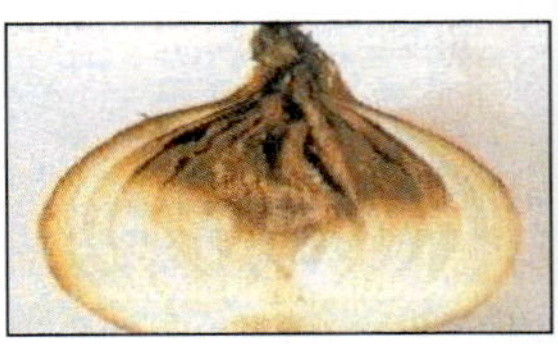

(c) Neck Rot
(https://www.rhs.org.uk

(d) Purple Blotch

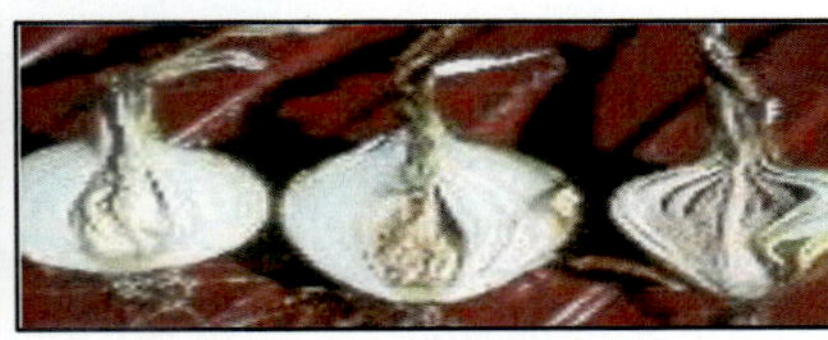

(e) Soft Rot

Postharvest Diseases of Onion (p. 136)

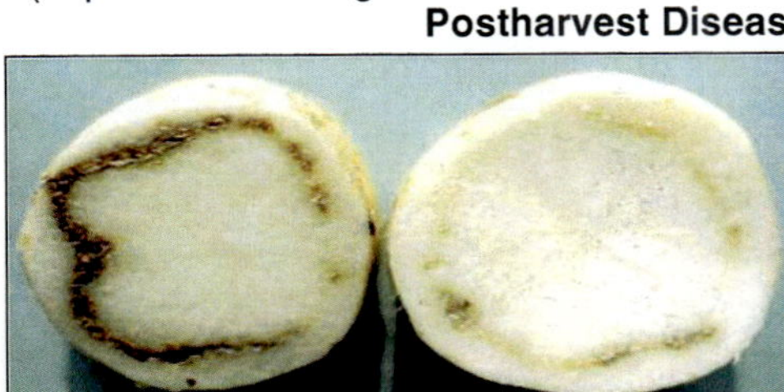

(a) Brown Rot
(http://www.cals.ncsu.edu and

(b) Charcoal Rot

(c) Dry Rot

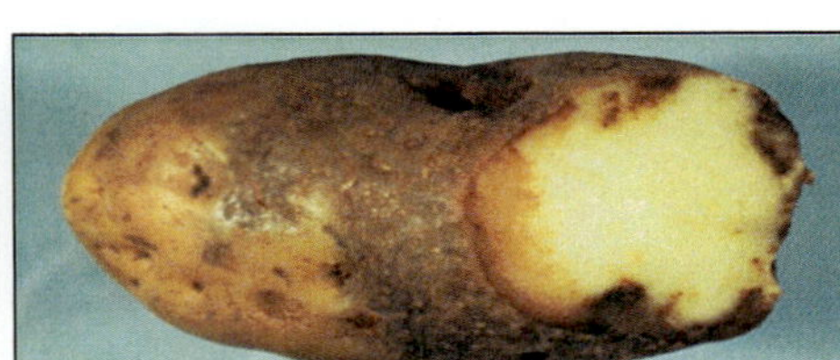

(d) Late Blight

Postharvest Diseases of Potato (p.138)

(a) *Alternaria* Rot

(b) *Anthracnose*

(c) Bacterial Soft Rot
(http://
agritech.tnau.ac.in)

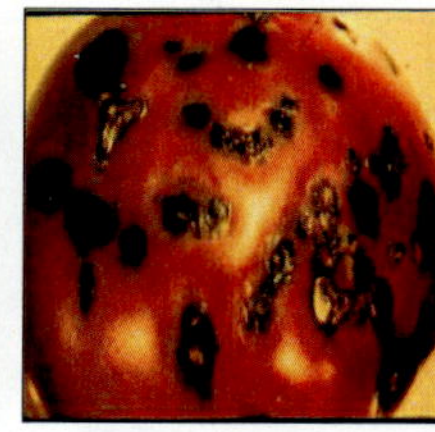

(d) Bacterial Spot
(http://
ipm.uconn.edu)

(e) Buck Eye Rot)

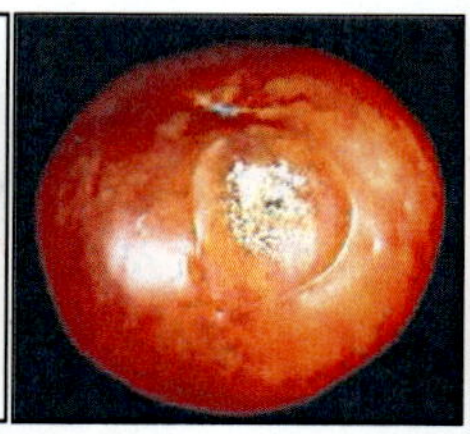

(f) *Fusarium* Rot

(g) Grey Mould Rot

(h) *Phoma* Rot

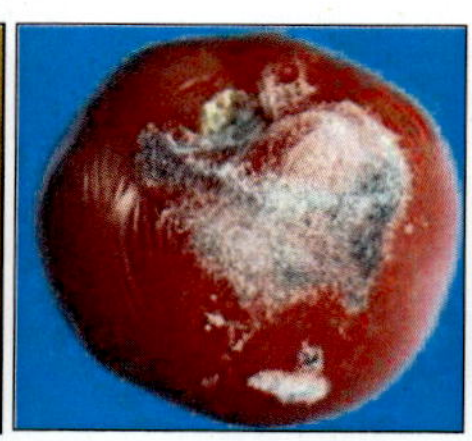

(i) *Rhizopus* Rot

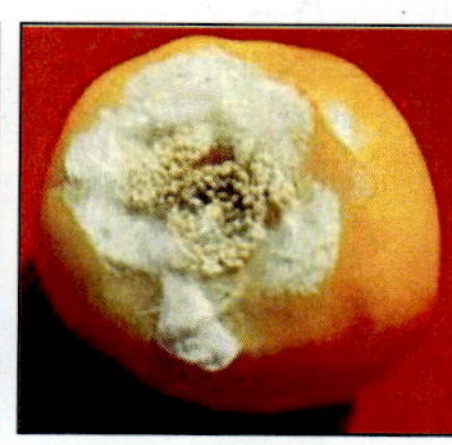

(j) *Sclerotium* Rot)

(k) Sour Rot)

Postharvest Fungal Diseases of Tomato (p. 142)

Telescopic Corrugated Fibre Board Container (p. 154)

Wooden Pallet Board at the Bottom with Stacked Corrugated Boxes in Pallet form for Unit Handling (p. 155)

Apples Placed on Moulded Trays and Packed in Carton (p. 165)

Apples Packed in Polystyrene Foam Sleeve and Packed in Open Top Carton (p. 165)

Shrink Wrapped Green Alphonso Mangoes after Ripening (p.188)

Shrink Wrapped Pomegranate
(a) Ambient Condition (Above) (b) Refrigerated Condition (Below) (p. 200)

Sapota Fruit Packed in Corrugated Fibre Board Boxes (p. 201)

Strawberry Packed in Plastic Punnet (p. 203)

Fruit Fly Damage in mango by *Bactrocera dorsalis (p. 214)*

Damage by Mango Stone Weevil (p. 215)

Burying of Infested Fruits (p. 226)

Bait Splash on Trunk (p. 226)